ON

MATTER AND ETHER.

ON

MATTER AND ETHER

OR

THE SECRET LAWS

OF PHYSICAL CHANGE.

BY

THOMAS RAWSON BIRKS, M.A.
RECTOR OF KELSHALL, HERTS
FORMERLY FELLOW OF TRINITY COLLEGE, CAMBRIDGE
AUTHOR OF "THE DIFFICULTIES OF BELIEF" &c.

Cambridge:
MACMILLAN AND CO.
AND 23, HENRIETTA STREET, COVENT GARDEN,
London
1862

Cambridge:
PRINTED BY C. J. CLAY, M.A.
AT THE UNIVERSITY PRESS.

PREFACE.

THE Theory unfolded in the present Essay, whatever may be its merits or defects, is not the hasty production of a moment. It is twenty-eight years since the first steps were taken in the line of inquiry, which has now at length assumed a consistent and connected form. Three main elements of the theory, the conclusion from the law of gravitation with regard to the nature of the atoms of matter, the constitution of chemical elements, as the first step in composition of these material monads, and the large part played by rotatory motion and centrifugal atomic force, in nearly all branches of physics, were then imperfectly traced out in connexion with various kinds of phenomena, and became settled convictions of my mind. A second step, after some little interval, was to discern the absolute need of admitting ether distinct from matter, as proved by the phenomena of light and electricity, and along with this, the consequent necessity for three distinct laws of central force, to explain their nature and mutual action. From these data alone consequences were traced out, at intervals of reflection, such as other engagements would allow, which seemed gradually to take a definite shape, and reveal the main classes of phenomena, known to science, as necessary results of such a constitution of the two primary elements. One difficulty, however, hindered me from ripening the theory into a definite and tangible form, and made me unwilling, in spite of the many interesting lines of speculation it opened, to offer it to the thoughts and criticism of men of science, until it had been overcome. I was long unable to conceive such a relation of the constants, required by the laws of force, as would satisfy the phenomena of light and of the cohesion of solids, and also make mechanical structure a direct and immediate result of chemical composition. It seemed needful to allow a wide interval between the chemical atom, the first result of composition, and the molecule, on which cohesion and solid structure depend. In a later review of the subject, by combining the known data, and adopting an inductive course, this difficulty, I believe, was finally removed. Its source was the assumption, on grounds of apparent simplicity, of the lowest possible powers, the inverse third and fourth, for the two unknown laws of force to be determined. By the course of reasoning unfolded in the earlier chapters of the Essay, this source of perplexity was removed, while the main features of the theory, as it had for many years been developed, remained unaltered. All recent discoveries, during the thirty years since I became possessed of some of its main outlines, seem to me to have only confirmed, by anticipation, its substantial validity and truth. With the strong hope that it will be found to supply the true key to many of the undisclosed mysteries of nature, which have hitherto baffled all attempts at harmonious and consistent explanation, I commit it now to the candid and thoughtful reflection of men of science, and to the blessing of Him in whom are hid all the treasures of wisdom and knowledge.

KELSHALL RECTORY,
Sept. 23, 1862.

CONTENTS.

ON MATTER AND ETHER,

OR THE

SECRET LAWS OF PHYSICAL CHANGE.

INTRODUCTION.

SINCE the discovery of the Law of Gravitation, by Sir Isaac Newton, an immense progress has been made in every branch of physical science. Chemistry and electricity were then in their infancy, and galvanism quite unknown. But the advance, of late years, has been rapid and continual. Secondary laws of high importance have been discovered, as in the Undulatory Theory of Light, researches on specific and radiant heat, electrical attraction and induction, atomic proportions, and the laws of crystallization. All the subtler influences of nature, light, heat, electricity, magnetism, chemical affinity, crystalline polarity, are found, more and more, to be intimately related to each other. Men of science feel themselves to be on the verge of some great discovery, but the key which can unlock these various secrets of nature has not yet been attained.

The theories which have been proposed, to explain separately some one class of these phenomena, are plainly insufficient. Thus electricity has been referred, sometimes to one, and sometimes to two electric fluids. The second hypothesis has been developed by Coulomb, Poisson, Whewell, Murphy, and other analysts. But the reasons why these two fluids should combine with matter, and nearly all their laws of combination, remain wholly unexplained; while the supposition itself, of two such fluids differing only by a positive and negative sign, is very remote from natural probability. Again, magnetism has been referred, by Ampère, to spiral systems of electric currents. But the needful postulates, that two elements of electricity attract each other, when they move in the same direction at right angles to the line of junction; that they repel with the same force, when they move opposite ways, and with half the force, when the motion is in the direction of their distance, have none of the simplicity of ultimate laws. No explanation at all is given, why electricity in motion should attract differently from its state of rest, or why currents in constant revolution should exist around the particles of a magnet. The theory may have its use as a landing-place in the ascent of science, but the true nature both of electricity and magnetism clearly remains still to be discovered.

Again, theories of heat have been constructed, by Fourier and others, with great analytical ability, on the hypothesis that caloric is a fluid, condensed around the molecules of matter, and radiating constantly from one part of it to another. But the later discoveries of the Polarization of Heat, and of its equivalence with mechanical force, have almost entirely overthrown the idea that it is a distinct and separate fluid. All these theories, of two electric fluids, a magnetic fluid, and a fluid of heat, are not only fragmentary, but divergent, and in their results contradictory to each other.

Several attempts have been made to propound some view of the constitution of matter, which may account for the various forms of molecular action. The theory of Boscovich was one of the earliest. Each particle was supposed to be a point, endued with attractive and repulsive force; the curve of force being such as to cross the axis several times, or to have several neutral distances, where the force changes from attraction to repulsion, the repulsion tending to infinity at one limit, and the attraction varying as the inverse square at the other. But a law of force with these conditions is so complex, and admits of so many arbitrary varieties, as to make the hypothesis highly improbable, and practically useless. The modern discoveries in optics, also, seem to exclude the notion that one kind of matter alone will account for all the varied phenomena of the universe.

The first rule of philosophical reasoning, laid down by Sir Isaac Newton, is to this effect: "No more causes of natural things are to be admitted, than such as are both true, and sufficient to explain their appearance." He says further, at the close of the work: "Whatever is not deduced from the phenomena is to be called an hypothesis; and hypotheses, whether metaphysical or physical, whether of occult qualities or mechanical, have no place in experimental philosophy."

The rule, thus stated, is clearly open to a very weighty objection. If the true cause is already known, all inquiry is superfluous. But if not known, how can it be the test

1—2

of a sound induction? Dr Whewell has observed, more accurately, that hypothesis is essential to the progress of science; that few discoveries have been made, till several hypotheses have been tried; and that the test of a sound induction is not the rejection of all hypothesis, but a readiness to invent those which shall be most promising, and to abandon those which have been proved insufficient. It is like the use of a bunch of keys in unlocking a door. He succeeds best who only tries the keys which have some apparent resemblance of size to the lock itself, and who, instead of trying to force the lock with a wrong key, lays it aside quickly, and tries another. There can be no more complete test of an hypothesis than its power to explain all the phenomena; and the word "true" in Newton's rule, may thus appear not only superfluous, but to involve a logical contradiction. The real meaning, however, of Newton, was probably this; that a cause already accepted for one class of phenomena, if it will also explain another class, is preferable to one which explains the latter alone. Thus gravity was a known fact with regard to bodies on the earth's surface; and since it would account also for the moon's revolution, and the orbits of the planets, it was to be preferred to any explanation which would apply to the latter only. The principle, thus understood, is the same which has been well called "the consilience of inductions," and seems to lead to the following principles or rules of inductive inquiry.

Axiom I. There are only two tests of the truth of any physical hypothesis; its fitness to account for all the phenomena, and its simplicity.

Axiom II. The simplest hypothesis, which offers any hope of explaining the facts, ought first to be tried; and more complex ones, only when the simpler has been tried, and proved to be insufficient by careful examination.

AXIOM III. The first step in the required proof of any hypothesis, is when it can be shewn to produce, by natural consequence, all, or nearly all, the same classes of phenomena, which a true theory is wanted to explain. Such an hypothesis may be called probable, and has a claim to fuller development.

AXIOM IV. An hypothesis is not only probable, but almost certainly true, when, on being developed, it yields a large variety of measurable results, which agree in quantity with the results of direct experiment.

AXIOM V. The simplest hypothesis is that which includes the smallest number of arbitrary postulates, such as distinct laws of force, constants of force, and constants of distance.

The first result of these simple axioms, when applied to the great problem of physical science, still unsolved, is to sweep away all the specific fluids, which have been conjecturally proposed, each for one limited class of phenomena; the two electricities, the fluid of heat, or caloric, and the magnetic fluid. The phenomena of optics seem to compel the admission of a luminous ether, besides matter, of immense elastic force. Until it has been shewn that this double admission, of ponderable matter, and of luminiferous ether, is insufficient to explain the phenomena, the recognition of any further varieties, or of fundamental diversities in matter itself, or of many distinct and unchangeable material substances, is opposed to the true laws of a sound induction. But before the inquiry can make any considerable progress, it will plainly be needful to form some clear and definite conception with regard to the real nature of both these kinds of substance, and the laws of mutual action which must be supposed to exist between them. Such will be the first object of the present investigation. It will then be endeavoured to trace out, in order, the main consequences of the fundamental hypothesis, and their correspondence with the known phenomena of physical change.

CHAPTER I.

ON MATTER AND ETHER.

1. *Every particle of matter attracts every other particle, with a force varying inversely as the square of the distance between them.*

This is the great discovery of Newton, which must form the natural starting point for every further advance in physical science. The history of Astronomy for two centuries has consisted mainly in the development of its results. But rich and fertile as its consequences, in one direction, have proved to be, those which flow from it in another have never been traced with sufficient clearness. When examined narrowly, it points to conclusions of great importance and scientific value.

2. *Every particle is either a mathematical point, or else contains such a point, as the true centre from which the attraction proceeds.*

By the fundamental law, the attractive power varies inversely as the square of the distance. But distances cannot be measured from a space of finite dimensions. They must be measured from some point only. Hence, if the attractive forces exerted by a particle on every other are determinate, then the point from which they emanate, or from which all the distances are reckoned, must be determinate also.

3. *There is no reason for admitting, in the particles of matter, a solid sphere of repulsion, enclosing the true centre of force, and distinct from it.*

First of all, such an assumption is very complex and arbitrary, and thus violates the first and second Axioms of inductive inquiry. A central force, emanating from a definite point or centre, is proved to exist. The hypothesis introduces these further elements; an arbitrary radius of the solid nucleus, an arbitrary shape of the atom, a surface of abrupt and infinite repulsion, a structure unalterably rigid, and also an arbitrary relation of the centre of force to the supposed solid nucleus. None of these things can be rightly assumed, while there is no clear necessity for the supposition.

Again, two of these assumptions are opposed to all the conclusions from experience. They would rest a scientific datum upon a popular impression, which exact science disproves. In compound bodies, repulsion always begins at some distance from the surface. A sudden change, then, from attraction to infinite repulsion, has no analogy in its favour. The most solid bodies, also, have always some elasticity, and a case of perfect rigidity is not to be found. The hypothesis, then, of a rigid, solid nucleus, not only finds no warrant in the law of gravitation itself, but is also opposed to all known analogy. Hence, by Axiom II. it is inadmissible, until it can be shewn that a simpler hypothesis, the natural result of the law alone, fails to satisfy the actual phenomena.

4. *The simplest view of matter, derived at once from the law of gravitation, is that it consists of monads, or move-*

able centres of force, unextended, but definite in position, which attract each other with a force varying inversely as the square of the distance between the centres.

This conception, of points that are centres of force, results plainly and unavoidably from the nature of the law of gravitation. Any further conception of the constitution of matter is an unproved addition. Also the conception, in this simplest form, involves no greater metaphysical difficulty than may be shewn to exist equally in every other conception of primitive or constituent atoms.

5. *The particles of matter, so constituted, could never coalesce with each other, and thus lose their individual being, or disappear.*

Let us suppose two particles to fall from rest in the line joining their centres. Unless all other particles were disposed with perfect symmetry in reference to this line, and were also at rest, there must be some lateral disturbance. But in this case they will describe ellipses round their centre of gravity, and after their nearest appulse, will recede again. Hence it is plainly impossible that any two particles, so formed, should coalesce together.

6. *The law of gravitation, in matter so constituted, will not alone account for the cohesion and solid structure of bodies.*

Let us assume one thousand millionth of the earth's radius ($=\frac{1}{4}$ in. or $6\frac{1}{3}$ millimetres nearly) for a linear unit. Assume further that a small sphere of this radius, and of the density of the earth, has its particles symmetrically placed, and 10^n for the number in its radius, or its mean linear density. The attraction of this sphere on an atom at its surface will be 10^9 less than the atom's weight. The attraction of atom on atom at that distance will be $\frac{4}{3}\pi.10^{3n+9}$

less than the weight of an atom, and at their own mean distance $\frac{4}{3}\pi.10^{n+9}$ less than the weight. Hence two atoms must be $64720\times10^{\frac{n}{2}}$ nearer to each other than this mean distance, that one may have a cohesive action on the other equal to its own weight, so as to retain it in permanent connection. There is plainly no arrangement of the particles which can satisfy this condition, and retain the least semblance of a solid structure. We must therefore look elsewhere for the explanation of cohesive force and solidity.

Two alternatives are possible. Either the law of gravitation must be modified for small distances; or there must be some other substance, distinct from common matter, on which the phenomena of cohesion depend.

7. *A self-repulsive ether, wholly distinct from common matter, also exists, and is diffused widely throughout all known space.*

The closing words of the *Principia* are like a prophecy, and shew in what direction the second main series of physical discovery must be attained.

"I might add something about a certain very subtle spirit, which pervades dense bodies and lies hid in them; by the power of which, bodies at very small distances attract each other, and when brought close together, cohere; and electrical bodies act at greater distances, attracting and repelling neighbouring bodies; and light is emitted, refracted, and reflected, and warms bodies; and sensation is excited, and the limbs of animals are moved at will, through vibrations of this spirit, propagated through the nerves to the brain, and from the brain to the muscles. But these things cannot be expounded in few words; nor is there extant a sufficient abundance of experiments, by which the laws of the activity of this spirit could be accurately determined."

In Sir Isaac Newton's *Optics*, the same thought appears in the modest form of queries.

"Is not heat conveyed through a vacuum by the vibrations of a much more subtile medium than air? Is not this medium the same by which light is refracted and reflected, and communicates heat to bodies, and is put into fits of easy reflexion and transmission? Do not hot bodies communicate their heat to cold ones by the vibrations of this medium? And is it not exceedingly more rare and subtile than the air, and exceedingly more elastic and active? And does it not readily pervade all bodies? And is it not, by its elastic force, expanded through all the heavens?"

Since the days of Newton, science has done much to supply the deficiency of experiment to which he alludes. The Undulatory Theory of Light, in displacing his own, has lent new evidence of the truth of his modest conjectures. In the hands of Young, Fresnel, Malus, Cauchy, Airy, Lloyd, and Stokes, it has come very near to astronomy in the singular and beautiful triumphs of its analysis, and the large variety of curious phenomena which have been explained. But its first postulate is the existence of an ether, such as the sagacity of Newton divined long ago, and indeed the same by which Huyghens, his contemporary, had already solved the phenomena of double refraction.

This ether must be diffused through all known space. For light is caused by its undulations, and this reaches us from stars immensely distant. Its elasticity, also, implies its diffusion in all directions, and not merely in the lines of the transmission of stellar light.

Again, this ether must be self-repulsive. If its particles were mutually attractive, they would evidently condense around centres, where there was any excess of density at first, and light could not pass from one of these condensing systems to another. Mutual repulsion, therefore, must plainly be one of its fundamental laws.

The denial of the existence of this ether, when confirmed by so many discoveries of modern times, though M. Comte in his Lectures ventures to style it a mark of superior wisdom, is a step backwards into the nonage of science. To recognise a cause, which is pointed out by many classes of phenomena converging together, is just as imperative a law of sound philosophy, as to reject and refuse every really superfluous element.

8. *No second fluid, of caloric, electricity, or magnetism, ought to be recognised, until it can be proved that the action and reaction of common matter and a luminous ether are incapable of supplying the required explanation.*

This results at once from the second Axiom. The simplest hypothesis may justly claim to be the first examined. That which recognises matter alone, is disproved by the phenomena of cohesion and of light. Next in order of simplicity is that which recognises both matter and a luminiferous and elastic ether, but no other fundamental variety. The ready adoption of so many hypothetical fluids, as have been often proposed, repels thoughtful students, and makes them ready to question the existence of any ethereal medium whatever, distinct from common or ponderable matter. The line of safe induction lies between these two extremes; and the most able analysts and experimentalists already tend to this middle view.

9. *The existence of matter and ether requires the ad-*

mission of three, and only three, laws of force for their mutual action.

First, matter acts on matter, and the law of its action is already known, being one of mutual attraction, inversely as the square of the distance.

Secondly, matter must act on ether, and ether on matter. This force must also be attractive, since otherwise the ether would not be condensed around the material atoms, nor give rise to cohesive affinity between neighbouring particles. Its exact law of force is hitherto unknown.

Thirdly, ether acts on ether. This force must be repulsive; since otherwise the ether would converge into patches, and not be diffused through space, as the transmission of light from the stars in every direction evidently proves it to be.

Fourthly, these two unknown laws cannot be the same with that of gravitation, or vary only as the inverse square. In this case all three forces would increase and diminish together, and cohesive attraction and ethereal repulsion would have the same relative amount at great as at small distances. There could thus be neither increase of cohesion, nor of resistance to pressure, however closely particles were packed together. These two forces, then, must follow some higher law than the inverse square.

10. *The ether of the universe greatly exceeds in quantity, or in the number of its atoms, the amount of ponderable matter.*

The force of cohesion, by the hypothesis, must depend on the presence of ether along with the material atoms. Hence, even in solids and liquids, the number of ether atoms must exceed that of the material atoms. But the planetary and stellar spaces are all filled with ether, and nearly void of matter. The space of the solar system half way to the nearest stars, exceeds the bulk of the sun more than 10^{22} to 1, or ten thousand trillion times. It does not seem likely that the mean distance of the monads of matter can be much less than that of the free ether in space. We may conclude that the number of ethereal is immensely great, compared with that of the material atoms.

11. *The mean distance of the particles of free ether must be less, and is probably far less, than one ten millionth of an inch.*

The violet rays of light make 60,000 undulations in one inch. With the limit assumed above, there would be only 167 ether particles in the length of a wave. In like manner 250 would be the number in the length of a red ray, and the difference about 80. It seems plain that, in a discontinuous medium of the kind suppposed, no vibration can be propagated to any distance, of which the length is not some multiple of the mean interval from atom to atom. But the black lines of the speculum divide it into a much greater number of different shades or kinds of light. Hence it seems to follow that the distance of the atoms must be less than $\frac{1}{10000000}$in. and may be very greatly less.

12. *The pressure of the ether on any surface must be immensely great.*

The pressure of common air, at the earth's surface, is nearly fifteen pounds per square inch. Now when two media are compared, the velocity of a vibration varies as the square root of the pressure, or modulus of the elasticity.

The velocity of sound, however, apart from its increase by the momentary heat, is only 916·3 feet per second, while that of light is 192,000 miles. The ratio is thus 1,106,360 to 1. The ethereal pressure, then, must be 1,224037,000000 times greater than that of atmospheric air, or $18\frac{1}{3}$ billions of pounds to the square inch. It is plain, then, how powerful must be the action of what are usually called the imponderable elements, or the mechanical forces, which light, heat and electricity, bring continually into play.

13. *The action of matter on ether must vary as the inverse cube or some higher law, and the repulsion of ether on ether, as the inverse fourth, or some higher integer power.*

First, the principle of the second Axiom requires us to assume integer, rather than fractional powers, for the two unknown laws of force, until decisive evidence to the contrary can be found, since the assumption is evidently far simpler. The nature of cohesion, also, and ethereal action, evidently requires inverse powers. But the action of matter on ether must increase more rapidly than gravitation for small distances, or else cohesion would not be limited to such distances. Hence the inverse cube is the lowest admissible power.

Again, the repulsive force of the ether must vary more rapidly than the affinity of matter for ether, since otherwise there would be no limit to the condensation, and attraction would increase the fastest with the density of the mass. The inverse fourth is the lowest power which can satisfy this first condition. Reasons, however, will presently appear for assuming still higher powers for the true laws of nature.

14. *The Three Laws of Force imply two independent*

constants, and two others result from the actual constitution of the ether in space, and of material bodies.

Because the action of ether on matter and ether follows two different laws, there must be some distance at which they are equal, or have a fixed proportion. The distance at which the attractive power of a monad of matter is half the repulsive power of a monad of ether, will be a First Constant. The distance at which the gravitation of matter to matter, and the repulsion of ether for ether, are equal, will be a Second, or Mean Constant. The distance at which the affinity of matter for ether is one half the attraction of matter for matter, will be a Third Constant.

The Second, or Mean Constant, is determined by the two others. It will be a mean proportional when the powers are equidistant, and will divide the ratio as $m-2 : n-m$, in other cases, where m is the index of affinity, and n of repulsion.

Again, the self-repulsion of ether implies its equable diffusion through space, apart from the condensing power of material atoms. There must therefore be a certain mean distance, on which its elasticity, and the propagation of light and radiant heat depend, and which will modify every kind of molecular action. This may be called the Ether Constant.

A cubic inch of water must contain a certain number of material monads, on which its weight depends. If these are all arranged symmetrically, at right angles, they will define a certain mean linear distance of the monads of matter, when the density is one. This distance may be called the Solid Constant. When we know the density of any body, and the number of monads in its chemical atom, the mean distance of those atoms will be deduced immediately from this solid constant.

15. *To determine some general relations among these constants.*

Let 10^{-m}, 10^{-n}, in inches, be the Ether and Solid Constants. The log. in grains, of the weight of a cubic inch of water, is 2·40219. The Earth's attraction, assuming its density to be 5·5, exceeds that of a cubic inch of water at one inch distance, log = 9·39790. But the action of one cubic inch on another must exceed that of one atom on another, as 10^{6n}. Hence $6n + 6·99571$, or $6n + 7$ nearly, will be the neg. log. for the force in grains, exercised by one material monad on another at one inch distance.

Again, the pressure of the air on a square inch is about 14·62 lb. = 102340 grs., log = 5·01005. But the ratio of the velocities of light and sound, apart from the heat of compression, has its log = 6·04390. The pressure in elastic fluids is as the square of the velocity of vibration. Hence the logarithm of the ethereal pressure, in grains per square inch, = 17·09785, or 125270 billions of grains.

The tenacity of a square inch bar of steel is about 134,000 lb., or nearly 10^9 in grains. The cohesive action in a substance, density = 1, with its monads evenly disposed, is probably less, or its log = $9 - x$, where the value of x depends on the chemical atom and structure, and the density.

If now we assume the Ether and Solid Constants to be equal, and the cohesion of steel to be nearly the same as of matter evenly disposed, density = 1, we have 17, 9, and $-(2n + 7)$, for the common logarithms of the repulsion, affinity, and gravitation, of 10^{2n} or 10^{2m} monads on a like number, on the surface of a square inch, and at a distance, each from each, of 10^{-m}. From this first comparison, then, it would seem to result that the Ether Constant, if not lower, must be nearly as low as the First Constant, since the monads of ether must have a longer mean distance, that the cohesion may equal the ethereal pressure, as deduced from the immense velocity of light compared with sound.

The thickness of a soap-bubble, before it bursts, has been proved to be only four ten-millionths of an inch. Hence n, the index for the Solid Constant, cannot be less than seven, and may be much greater. In like manner, it follows from the waves of light, that the mean distance of the ether monads must be less than one ten-millionth of an inch, or m greater than 7. But other facts lead to the inference that both of these indices have a much higher value, and that the mean distance of the monads of ether and atoms of matter is very considerably lower in order of real magnitude.

CHAPTER II.

THE LAWS OF AFFINITY AND REPULSION.

16. *To determine the general relations of Matter and Ether, under laws of Affinity and Repulsion, following a higher power of the distance than the Inverse Square.*

Let an atom of matter be placed in an ocean of free ether, and attract it by the law of the inverse cube, or by some higher power. It will be nearer to one monad of ether than to the others, or will become so by their motion, and will therefore be more strongly attracted by it, and attract it in return. The other monads, repelling the ether, and attracting the matter, will impress on the two a rotatory motion. But the centrifugal force produced by their mutual action, or that of the surrounding monads, will not, with the inverse cube or any higher power, be equal to the increase of the direct force. They will thus describe a rapidly decreasing spiral, in which the line to the centre is finite. Hence the two atoms must coalesce with an infinite force, and then remain inseparable.

This process of elective affinity, if the ether particles are in great excess, will continue until every monad of matter has combined with one of ether, so as to form a double or conjugate atom. This dual atom will be neutral to other dual atoms at the distance of the First Constant, where the one repulsion will balance the two affinities; and will be neutral to free ether atoms at a slightly greater distance, depending on the two laws of force. There will be a strong attraction at greater distances, and a still stronger repulsion at less distances.

All further coalescence of these double atoms with each other, or with the monads of uncombined ether, will be impossible. For when the distance is less than the First Constant, the repulsion will be greater than the sum of the two equal affinities, and at half the distance be at least twice as great; or if the indices of the powers differ by two, four, or six, it will be four, sixteen, or sixty-four times greater. Hence we shall have two kinds of centres of force only. The first consists of double monads, of matter combined with ether, following a double law of force, with a positive and a negative term. The second consists of free or uncombined ether, obeying its own simple law of self-repulsion.

The further relations of the dual monads, which we may call material atoms or units, to each other and to uncombined ether, will depend on the relation between the First and the Ether Constant. If the First be much less than the Ether Constant, or mean distance of the free ether monads, then each particle of matter will condense around it a series of ether particles up to and beyond the limit of that mean distance, so as to form an ether atmosphere. The constitution of the central part of this atmosphere will depend almost entirely on the law of central force, and the outmost strata alone will be much modified by the compressing force of the free ether. But if the First Constant equals the Ether Constant, only the nearest ether monads will experience a very sensible compression by the central force, and the others will be very slightly compressed, because the attracting power of the centre will be only a small fraction of the pressure from the nearest ether monads. In this case each particle of matter may have a single, double or triple line of attached ether monads, but not an ether atmosphere of dimensions much larger than the distance of its component monads. Again, if the First Constant were much larger than the Ether Constant, the cohesive force would bear a very small proportion to the repulsion of the ether, and a solid and coherent structure of bodies would be rendered impossible. Our choice is thus restricted to the two former alternatives alone.

17. *To determine the condensation of ether, apart from external pressure, with different assumed laws of Affinity and Repulsion.*

Let m, n denote the indices of affinity and repulsion, x any distance from the centre of force, and r the linear density, or number of ether monads in the unit of distance. Taking an infinitesimal cylinder of unlimited length, ar^3dx will be the differential of the mass, and $ar^3x^{-m}dx$ that of the force acting on it from the centre. But the pressure is produced by ar^2 monads, each at distance $\frac{1}{r}$ from the next, or having a repulsion as r^n. The differential of this pressure must be equal to the total action of the central force on the differential of the mass.

Hence $d.ar^{n+2} = (n+2)\, ar^{n+1}\, dr = ar^3x^{-m}dx$,

and, dividing by r^3 and integrating

$$(n+2)\,(m-1)\, r^{n-1} = (n-1)\, x^{1-m} + c.$$

Or, varying the form of the constant,

$$r = c\sqrt[n-1]{a^{n-1} + x^{1-m}},$$

where ca is the density of the free ether, apart from the attraction.

Neglecting the external pressure,

$$r = cx^{\frac{1-m}{n-1}}; \therefore r^3 = c^3x^{\frac{3-3m}{n-1}}.$$

The mass of a sphere, whose density follows this law, will be $\frac{4}{3}\pi.\frac{n-1}{n-m}x^{\frac{3(n-m)}{n-1}}$. And hence the cohesive action of two such atoms on each other, if nearly the same as when their atmospheres are condensed at the centres, will have for its negative index $m' = m - \frac{3(n-m)}{n-1}$. To satisfy the conditions, this force must vary by a higher law than the inverse square. But if $m=3$, $n=4$, $m'=2$, and if $m=3$, $n=5$, $m'=1\frac{1}{2}$. Hence, apart from external pressure, and in the case of matter surrounded by ether atmospheres, the law of the inverse cube fails to satisfy the rapid change of cohesive force that results from the phenomena.

If $m=4$, $n=5$, $m'=3\frac{1}{4}$; and if $m=4$, $n=6$, $m'=2\frac{4}{5}$; if $m=4$, $n=8$, $m'=2\frac{2}{7}$. Hence, in the same case, it may still be doubtful whether the increase and decrease of cohesive force would be sufficiently rapid, even with the law of the inverse fourth power.

If $m=6$, $n=8$, $m'=5\frac{1}{7}$; $m=6$, $n=10$, $m'=4\frac{2}{3}$; $m=6$, $n=12$, $m'=4\frac{4}{11}$; and in all these cases the actual law for the composite action is higher than the fourth, so as to satisfy the probable conditions, but exceeds it less, as the index of repulsion is increased.

18. *To determine, by a first approximation, the arrangement of ether atoms along a single radius, neglecting, as before, the action of the free ether.*

A single atom will plainly be in equilibrio at the neutral distance for ether, which exceeds that for matter in the ratio $\sqrt[n-m]{2}$. Taking this for the unit, we may, in a first approximation, suppose the attraction of the central matter to be balanced by the repulsion of the one ether monad, which lies nearest it towards the centre, and that the other ether monads on the two sides nearly compensate each other. Let a_0^n, a_1^n, a_2^n, a_3^n, &c. be the successive total distances of the ether monads from the centre; also $n = \alpha(n-m)$ and $a_r^{n-m} = b_r$. Then $b_r^{\alpha} - b_r^{\alpha-1} = b_{r-1}^{\alpha}$ and $b_r = b_{r-1} + \frac{1}{\alpha}$ nearly. Thus $\left(1 + r\frac{n-m}{n}\right)^{\frac{n}{n-m}}$, will nearly represent the successive distances of the ether monads, and the number of those in the distance l will be

$$\frac{n}{n-m} l^{\frac{n-m}{n}}, \text{ nearly.}$$

Applying this result to the case of spherical atmospheres of different radii, and putting μ for the number of monads, or the mass of condensed ether.

$m = 3,\ n = 4,\ \mu = 85\tfrac{1}{3}\pi r^{\frac{3}{4}}.$ $m = 6,\ n = 8,\ \mu = 85\tfrac{1}{3}\pi r^{\frac{3}{4}}.$

$m = 4,\ n = 5,\ \mu = 166\tfrac{2}{3}\pi r^{\frac{3}{5}}.$ $m = 6,\ n = 10,\ \mu = 20\tfrac{5}{6}\pi r^{\frac{6}{5}}.$

$m = 4,\ n = 6,\ \mu = 36\pi r.$ $m = 6,\ n = 12,\ \mu = 10\tfrac{2}{3}\pi r^{\frac{3}{2}}.$

All these relations will be essentially modified, if the ether constant be nearly or quite as small as the first constant, or the pressure of the free ether has a sensible proportion to the forces which balance each other at the neutral distance.

19. *To deduce the probable Law of Affinity from the comparison of weight and cohesive force.*

The tenacity of a square inch of steel, the highest known, is about 134,000 pounds, or nearly 10^9 in grains. But the gravitation of one cubic inch towards another, density 1, at one inch distance, is 10^{-7} in grains nearly. Hence the neg. log. for atom on atom at that distance, is $6n+7$, at the distance 10^{-n} it is $4n+7$, and for 10^{2n} atoms, or the action of each stratum on the next, $2n+7$. Assuming the former to represent the cohesive force of two adjacent strata in the uniform arrangement, this exceeds gravity as $10^{2n+16} : 1$. Supposing the tenacity to result entirely from the mutual force of the particles nearest to a plane section, and f to be the index of affinity, then $2n+16-(f-2)n$ will be the logarithm of cohesion ÷ gravity, at one inch distance. But general experience, and the special experiments of Mr Baily on the earth's density, tend to shew that cohesive force is small, if not quite insensible, even compared with the general force of gravitation (which is very small itself), at one inch distance. Hence we must have $(f-2)n > 2n+16$, or $(f-4)n > 16$. This seems to prove that cohesion must follow even a higher law than the inverse fourth power, and can only agree with the inverse fifth, if n is > 16, or the mean distance of the material particles not greater than 100 trillionths of an inch.

The difference between the cohesion of steel, and of the hypothetical arrangement of monads in matter of density one, may a little modify these values. But since the disproportion is probably far less than 10^{16}, it will not affect the main conclusion, that no inverse power lower than the fifth can satisfy the conditions for the law of affinity.

Since the law of gravitation is the inverse square, it seems more simple to assume even powers for the other laws of affinity and repulsion. Hence the Inverse Sixth is the simplest law of affinity, which satisfies the previous condition. In this case, if $n > 8$, or the mean distance of the material particles is less than one hundred millionth, the great superiority of cohesive force above gravitation at the Solid Constant, and its insensible amount at one inch distance can both be fully satisfied.

20. *To determine the probable Law of Repulsion.*

The Law of Affinity having now been assumed to be the Inverse Sixth, as the simplest which agrees with the conditions, the Law of Repulsion must be still higher, and simplicity leads us to infer that it is also some even power. The inverse eighth, tenth, and twelfth, have each some claim to be the simplest. The first of these is the lowest even power, after the sixth. The inverse tenth makes the intervals equal between each pair of indices, since $10-6=4=6-2$. Again, the inverse twelfth has an index the product of the two others, being the sixth power of the law of gravity, and the duplicate of the law of affinity. Its adoption makes the sphere of the three successive forces more distinct, or the repulsion more marked in the first, and the predominance of cohesion in the second sphere, and the relation of the powers is both simpler in itself, and with reference to calculation. On this view the three indices r^2, $r^{3.2}$, $r^{2.3.2}$ have the simplest relation to each other, as products. The second is the triplicate of the first, and the third a duplicate of the second. The main features, however, of the general theory will be the same, if a further and more exact analysis were to indicate some other law than precisely the inverse twelfth as the true constitution of ether.

Let us now, to fix our ideas, assume that the Ether Constant is 10^{-18} or one trillionth of an inch, and the Solid Constant ten times greater. In this case 10^{17} is the ethereal pressure in grains at a square inch, and 10^{9+x}, the cohesive force on a square inch, at a distance ten times greater. But when the distance is lessened ten times, the increase of the particles will be 10^2, and the action of each increased 10^6, so that 10^{17+x} will be the cohesive action at the distance of the Ether Constant. On this supposition, according to the value of x, the first constant will be greater or less; but only in a small proportion, than the Ether Constant, and the solid constant greater than one or the other in a nearly tenfold ratio. We seem thus led naturally to the conclusion, from the great velocity of light, that both the Ether and Solid Constants, depending on the actual constitution of the universe, must approach, in their actual values, to the magnitude of the first and lowest of those three constants, which result from a direct comparison of the laws of force themselves.

CHAPTER III.

ON THE GENERAL FORMS OF MATTER.

21. THE Four Elements of the ancients, though displaced by modern chemistry, still retain in reality an important scientific meaning. The element of Fire finds its counterpart in the doctrine of the imponderables, or light, caloric, and electricity. Air, Water, and Earth, again, represent the three distinct forms under one or other of which all known substances must be classed, as gases, fluids, or solids.

On the other hand, the sixty species of unresolved substance, which are now called elements, represent evidently an imperfect and provisional stage of chemical analysis. They have no natural claim to be viewed as distinct and inconvertible kinds of substance, wholly incapable of transmutation. They are specific distinctions, while those of the Four Elements, rightly expounded, are generic, and thus rank perhaps even higher in their scientific value.

Our first impressions of matter are derived from Solids, and their resistance to the touch, or pressure of the hand. Hence the nature of matter has often been defined by extension and solidity. But the progress of science has shewn the error of such a definition. The sensation of solidity is evidently compound, and arises from a repulsive force exercised along a well defined surface. This repulsion begins before actual contact, and appears to be a rapidly decreasing power, which emanates from the outer particles of the resisting substance. When it is melted or vaporized, the substance remains, but its law of force is altered, and the sensation of solidity disappears.

Again, the impenetrability of matter, in the popular sense, is disproved by the facts of chemical combination. "We may cast into potassium oxygen, atom for atom, and again both oxygen and hydrogen in a twofold number of atoms; and yet, with all these additions, the matter shall become less and less in bulk, till it is not two-thirds of its original volume. A space which would contain 2800 atoms, including 700 of potassium, is found to be filled by 430 of potassium alone." (Faraday.) Hence the impenetrability of matter is illusive, and the term does not correctly describe its real nature.

One of the first requisites in a theory of the constitution of matter, is to supply a general explanation of solidity, fluidity, and gaseous expansion, the three fundamental states in one or other of which all known substances are found to exist.

22. *To explain generally the cohesion of Solid Bodies.*

Let a number of atoms, such as have been already defined, be diffused through an ocean of ether. If the First Constant be much lower than the Ether Constant, each will condense around it, separately, an atmosphere of ether monads, in large numbers. If slightly lower, equal, or slightly superior, it will still attract and condense the nearest ether monads, detaining them by a very powerful affinity. In either case there may be three degrees of density.

First, let the mean distance of the material atoms be greater than the third constant. In this case their mutual influence will depend almost wholly on the law of universal gravitation. They will condense gradually towards the centres of gravity of the denser portions; while the local pressure thus occasioned will drive away part of the ether, and produce a repulsive action on the less attracted portions.

Secondly, let the mean distance approach or fall below the third constant. The affinity of the nearest atoms will now be greater than the force of gravitation, and will increase rapidly. There will be thus a rapid condensation, unequally distributed, which will tend to a granular or mottled texture of the general mass. The more rapid condensation of some parts, by producing *vis viva*, will increase the repulsion of other portions; and while some parts approach to a solid or fluid state, in others that repulsion will predominate, or there will be a resemblance to a gaseous constitution.

Thirdly, let the mean distance of the material atoms be a small multiple of the first constant. Their own mutual action being double that which they exercise on the free ether monads, the nearest will tend to approach still nearer, till they reach the neutral distance, at which they neither attract nor repel each other; or one still smaller, where the excess of the repulsive force is equal to the mean ethereal pressure. At this distance any number of them may assume a settled form of equilibrium, in which the nearest distances are slightly less, but the diagonals greater, than this neutral distance. When these compound atoms are wide apart, their action on each other will not differ sensibly in its character from that of simple atoms. But when their distance is small, or they are in rapid rotation, their mutual action will depend on their shape, the number of their components, and the speed of their rotation, and polar relations must appear.

Each of these combinations of monads, or compound atoms, must have at least one tier or stratum of ether monads condensed upon it, or adhering to its outer surface. When they approach each other, there will thus be a strong repulsion at each surface; and this will tend to produce a rotation of the atoms round some axis of revolution.

If the atoms are distant, the cohesive force feeble, and the *vis viva* great, the motion will be around the axis of greatest moment. But when the density is greater, and the cohesive force in consequence is large, the atoms will be confined in the direction of the greatest moment, and will either revolve round a longer axis, or may simply oscillate without revolution. Many such molecules, joined together by this mutual polarity, will have all the qualities which belong to solid bodies.

23. *To determine further the relations between the fixed constants, in connection with the phenomena of gravitation and cohesion.*

Let m, n be the neg. logs. of the ether and solid constants, and a, $a-r$, $a-2\frac{1}{2}r$, those of the first, second, and third constants. Also l the negative logarithm of the cohesive limit, where the action of two atoms on the ether belonging to each other is together equal to their mutual gravitation.

From the velocity of light, we obtain 10^{17} for the ethereal pressure in grains, at the distance of the ether constant. Therefore $2m-17$ is neg. log. for one pair of monads at that distance, and $2m-17+12m-12(a-r)$ the same at the second constant. But $6n+7$ is neg. log. for attraction of atom on atom at one inch, and $6n+7-2(a-r)$ th same at the second constant.

By the definition of second constant, these are equal;

$$\therefore\ 14m=6n+24+10(a-r) \text{ or } 7m=5(a-r)+3n+12.$$

Again, the cohesion on a square inch, at the distance of the solid constant, is 10^{9+x}, being 10^9 in the case of steel.

But the attraction of one cubic inch on another at that dist. has neg. log. $=7$, that of atom on atom, $6n+7$, that of 10^{2n} atoms at dist. 10^{-n}, $2n+7$;

$$\therefore\ n-\frac{1}{4}(2n+7+9+x)=\frac{1}{2}n-4-\frac{1}{4}x$$

is the neg. log. of the cohesive limit, or the distance where cohesion and general gravitation are equal.

In the case of a condensed ether atmosphere,

$$l=a-\frac{5}{2}r-\frac{3}{4}(a-m)-\frac{1}{2}.$$

But when the first and ether constant are near together, it is more simply,

$$a-2\tfrac{1}{2}r=\frac{1}{2}n-4-\frac{1}{4}x.$$

Combining this with the former, since

$$5(a-r)=3a+2(a-\frac{5}{2}r),$$

we have

$$7m=3a+n-8-\frac{1}{2}x+3n+12=3a+4n+4-\frac{1}{2}x,$$

a general relation between the first, the ether and solid constants.

It results from this equation, that the solid must be larger than the ether constant, nearly in the same ratio as this exceeds the first constant. But the phenomena of density, in which the mechanical properties of bodies depend so directly on their chemical nature, make it highly probable that the solid constant is only a small multiple of the first constant. Hence the ether constant must have a value of the same order lying between them, unless it were to be even less than the first constant.

The supposition, then, of a vast disproportion between the first constant or neutral distance, and the ether constant, or of a large atmosphere of condensed ether round each atom of matter, seems excluded by the conditions; and the other alternative established, of one, two, or three tiers only of discrete ether monads, at finite intervals, being grouped around them.

24. *To determine the probable relation between the first constant, or neutral distance, and the solid constant, or mean distance of the atoms in a body of density one, from the general character of the known facts with regard to density and pressure.*

On the present theory, the first constant is that distance, below which there is a growing increase of repulsive force, varying as the inverse twelfth power. Now the whole ethereal pressure has been already found to be 10^{17} in grains per square inch. But the weight of a column of one square inch from the surface to the centre of the earth is only $10^{11.24}$, or six hundred thousand times less. Hence the condensation which it can cause, beyond that already produced on the nearest monads under the pressure of the free ether, must be very small. And since platinum is nearly four times heavier than the mean density of the earth, we have a presumption that it approaches to the maximum density, consistent with the value of the first constant, the mass of the earth, and the elasticity of the free ether surrounding our planet. We may thus assume a density $= 27$ as a near approach to the limit, or that which would result, if all the monads were ranged symmetrically at their neutral distance, diminished by the constant pressure of the ether, with no ether interposed, or chemical structure. It follows at once that the solid constant, or the mean distance in a substance of the density of water, is just about three times the neutral distance or first constant. The lightest solid is about four times lighter than water. It follows that the greatest linear disproportion of density, consistent with a solid structure, is as $\sqrt[3]{108}$, or $4\frac{3}{4}$ to 1, from cork to a density greater than of platinum, where chemical structure ceases at the extreme limit of terrestrial compression.

But if the solid constant is only about three times the neutral distance, and the ether constant lies between them, the problem assumes a much more definite form. In the equation of condition, we may take $n = a - \frac{1}{2}$, that is, the solid $= \sqrt{10} \times$ first constant, and we have $7m = 7a + 2 - \frac{1}{2}x$, from which it seems to result that the ether constant is either as small, or rather smaller, than the first constant. For the value of x depends on the possible increase of cohesion in a symmetrical arrangement of the material monads beyond its experimental amount in steel or iron, and unless this excess were as 10^4, which seems very unlikely, the equation would make $m > a$. On the other hand, no allowance has been made for the probable increase in the velocity of light, like that of sound, for the local compression of the vibrating ether, which might be considerable.

It may thus be inferred, finally, that the first and ether constants are sufficiently near to allow us to assume the larger to be double, or less than double, the other, and the solid constant to be only three or four times the first constant. If we assume this last, for convenience, to be one trillionth of an inch, and the others to be between this limit and ten trillionths, we shall satisfy, perhaps, all the known conditions, and render our conceptions more definite.

25. *To examine the general effects of atomic motion in modifying the cohesion.*

Let a compound atom, with its attached ether monads, be conceived to revolve round any axis. The centrifugal force will throw out the particles near the equator, whether of matter or ether, in proportion to the rapidity of rotation. There will thus be an increase of repulsion to the ether in the plane of the equator, but also an increased attraction in the line of the axis. For the *vis viva* created by the repulsion at the equator must diffuse itself through the medium, and occasion a greater pressure of the ether in the line of the two poles, while the partial expansion of the radius allows a closer approach for equilibrium. Thus, if the square of an octahedron were enlarged, the points must approach it, the side remaining constant.

In simple atoms, the centre has no moment of rotation, the attached ether can have only a slight polarity, and the axis of revolution will vary with ease. But in compound atoms, the material monads revolve round the axis, and will be thrown out beyond their natural distance. Such a revolving atom, in proportion to its *vis viva* of rotation, will be stable, and resist any separation of its components, or change of the axis of its rotation.

Let two such atoms revolve round parallel axes with opposite poles on the same side. There will be repulsion at the equator from the radial centrifugal force; but the tangential motions will agree, and leave them neutral to each other. But if like poles are on the same side, the tangential motions will be opposite. The impact of the tangential ether will make the one move round the other in the same direction; or transfer part of the moment of rotation from the separate atoms, and their two axes, to the axis through their centre of gravity, midway between them.

Again, if the axes are parallel, and similar poles on the same side, but the line of the centres oblique to that of the axes, the descending equator of the first will depress the south pole of the second, and the ascending equator of the second, so that both will tend to a new position, with an axis of rotation at right angles to the line of the centres. This new moment, combining with the first, will incline the poles to a diagonal position, each towards the equator of the other atom; and their repulsion, from the opposite motions of the attached ether, will cause a further change, till the axes of rotation are at right angles, not only to the line of the centres, but also to each other.

26. *To trace the cohesive structure of solids under these conditions.*

Let us suppose a cubic space, filled with material monads, at a mean distance, not many times greater than the first constant, assumed to be one trillionth of an inch. The affinity is immensely superior to the force of general gravitation, and will create a general viscosity or mutual adhesion of the whole mass. At the same time, the greater part of the space will be occupied by uncombined ether atoms, which will transmit on every side the general ethereal pressure. The affinity, varying as the inverse sixth power, will draw the nearest monads of matter strongly together, and in their approach they will rotate round each other in some plane. They will thus become partly grouped in revolving doublets or triplets, and the rest will continue at first to be separate monads. The repulsive force of the rotation will tend to equalize the distances, and counteract the direct force of mutual affinity. But it will operate in the plane of each equator, and an opposite force of relative attraction must be manifested in the line of every axis. Hence the nearest particles, repelled by their equators, will tend to combine by their poles. Their mutual affinity, in this line, is not counteracted by centrifugal force, and their appulse producing a new moment of rotation, they will combine, and in combining, revolve round a new axis of greatest moment.

The first result, then, will be to combine the material monads into revolving planes or cycles of three, four, five, or more monads, disposed circularly at right angles to their axis of rotation; and the second, to combine two or more of these revolving planes around the same axis. But when the *vis viva*, generated by these combinations, has been mainly absorbed in permanent rotatory motions, all the atoms will have become strongly polar to each other. If this polarity be sufficiently powerful, it will change their rotation from the axis of greatest to one of least moment, which will be lengthwise, and allow the poles to retain their closest appulse to neighbouring atoms. Thus every compound atom, in virtue of its shape and revolution, will have become an elementary magnet, and assume and maintain a polar and nearly fixed position with regard to all the neighbouring atoms. They will thus constitute a solid and coherent structure.

Again, the phenomena of light, in transparent bodies, and the greater amount of the ethereal pressure, compared with cohesive force, in these and all others, prove that ether is diffused throughout their whole texture, so that the general external pressure penetrates the whole. But this is excluded from the chemical atoms, by their assumed structure. It follows that the pressure of the free ether, propagated through the body, is that which isolates the chemical atoms from each other, so that at least a single range of ether particles exists between them. Each chemical atom will thus have its components forced closer together by the whole amount of the ethereal pressure, and the cohesion will be mainly through the ether particles, interposed between them. In the densest solids the proportion of interposed ether will be the least, and in the lightest solids the greatest; while the cohesion, and elasticity or tenacity, will depend on the peculiarity of the atomic structure.

27. *To explain, by the present hypothesis, the liquefaction of solids.*

Sir H. Davy remarked, long ago: "It seems possible to account for all the phenomena of heat, if it be supposed that, in solids, the particles are in a constant state of vibration, those of the hottest bodies moving with the greatest velocity; and that in liquids and elastic fluids, besides the vibratory motion, the particles move round their own axes with different velocities." The more recent discoveries on the mechanical equivalence of heat confirm this general suggestion, and prove the correctness of Bacon's early conclusion, that heat is only a form or effect of atomic motion.

Let us suppose that vibrations, propagated through the ether, produce oscillations in the compound solid atoms, while retained by polarity in their own places, around their mean position. This increase of motion will produce an increase of repulsion, by which the atoms will recede from each other, and the body expand. By this same expansion, the polar force, which recalls them to their mean position, will be diminished. When the expansion and oscillation have reached a certain limit, the polarity will be too weak to destroy the angular velocity, and the particle will continue to revolve to another polar position. This revolution, at first, will be unequal and intermittent. But if the momentum, or collective *vis viva* increases, it will become nearly uniform. The former case will be one of viscous, and the other of more complete fluidity. Each revolution must be very rapid, from the greatness of the cohesive force at these small distances, and the attraction of a particle in any direction will be the mean of its values in every angular position. This equality of cohesion and repulsion in all directions constitutes the definition of a perfect fluid.

28. *To explain the evaporation of fluids.*

The particles of fluids must be in a state of rapid revolution round their own centres. The density will then be determined by the balance between affinity and external pressure on one side, and centrifugal force and ethereal repulsion on the other. A rise of temperature will increase the velocity, and with it, the centrifugal force; and part of the attached ether, being thus set free, will increase the repulsion. The fluid will therefore expand, and the particles recede.

Up to a certain limit, the cohesive force, or the excess of the central attraction over central repulsion, will increase by distance, and a certain amount of *vis viva*, or sensible motion, is absorbed by the increase of mean distance. Beyond this limit an increase of centrifugal force diminishes the cohesion. The repellent will then prevail over the attractive forces, or the cohesion and pressure, and the body will begin to pass into the gaseous form.

Let r be the radius of each revolving atom, measured from its centre of gravity to its outmost component atoms, and s the distance of their outmost atoms from each other, when they approach nearest in their revolution. The cohesive force will result mainly from the mutual action of these outer atoms, during a part of their rapid rotation, and the excess of their attraction over their repulsion. In their equilibrium, these are equal, and for maximum cohesion $12x^{-13} = 6x^{-7}$ or $x = \sqrt[6]{2} = 1{\cdot}12246$, or the mean distance is increased in the ratio of $r + \frac{1}{2}s : r + {\cdot}56123s$, or nearly as $2r + s : 2r + s + \frac{1}{8}s$. Beyond this limit, the cohesive force will only be lessened by the further expansion, and the liquid must therefore assume the gaseous form.

The expansion of water, from its maximum density to its boiling point, is nearly $1{\cdot}403061$, or in linear density $1{\cdot}013919$. Hence, by the formula, $2r + s = 8{\cdot}8s$, and $r = 3{\cdot}9s$ nearly. It would result, from this explication, that the distance of the nearest points of the revolving atoms is about $\frac{1}{4}$ of their radius, or one eighth of their diameter; though, of course, many other elements would have to be included in a more exact solution, which must depend on a correct view of the various chemical elements, and the mode of combination of the aqueous atoms.

29. *To explain generally a fourth condition of matter, which is neither solid, liquid, nor gaseous, but igneous or ethereal.*

A gas is defined by its elasticity, or its constant tendency to expand, when not restrained by external pressure. But the consequences of this definition at the upper limit of the atmosphere have never been clearly traced, because the fundamental laws of matter and ether have been unknown.

When we rise from the earth, the density of air decreases, heat is absorbed, the temperature becomes lower, and the elasticity grows more feeble. The limit of these changes must be a state in which elasticity ceases, latent heat is very great, and the temperature proportionally low. These facts admit an easy interpretation on the present theory.

In a gas under pressure, the centrifugal force, from the relative motions of the separate chemical atoms, with the repulsion of the ether, exceeds the cohesive force. If the pressure be withdrawn, the particles must recede further, and in receding, *vis viva* is absorbed, and their motion must become feebler. The velocity being the same, the centrifugal force diminishes inversely as the mean distance, while by the same expansion, the velocity itself must be lessened, till it almost disappears. The motion of translation, first, and then of rotation, will gradually cease. Thus the sensible temperature, which is measured by this motion, will decline, and approach to an absolute zero, but the latent heat will increase, and reach its maximum. For by the separation of the particles, and their approach to a state of rest, each will exercise its attracting and condensing power to the utmost on the surrounding ether, extending to those distances where gravitation begins to be stronger than cohesive affinity.

Such a state of matter cannot be called gaseous, since all the defining properties of a gas have disappeared. Elasticity and expansiveness are replaced by inelastic repose. The want of intestine motion, either of translation or rotation, will make it resemble solids rather than liquids, while in density it occupies the other extreme; and being saturated with ether, it will approach to the properties of the free ether which it adjoins.

There will thus, according to the present theory of the laws of matter, be more truth than has latterly been recognized in the old arrangement of the four elements, which placed a fourth region of fire above the solid, liquid, and gaseous constituents of our globe. In fact, above the region where the air, though greatly rarified, is still elastic, there must be a still higher stratum where elasticity has wholly ceased, and where the particles of matter, being very widely separated, condense around them the largest amount of ether. All sensible heat, in the collision or oscillation of neighbouring atoms of matter, will thus have disappeared; but latent heat, in the quantity of condensed ether, or repulsive force ready to be developed on the renewed approach of the atoms, will have reached its maximum; and may be capable of producing the most splendid igneous phenomena, like the northern lights, or tropical thunderstorms.

30. *To explain the mutual relations of the solid, liquid, gaseous and igneous forms of matter.*

The laws of affinity and repulsion imply a first constant, or minimum distance, at which two atoms of matter, each charged with one of ether, will be neutral to each other, repelling at less, and attracting each other at greater distances. Chemical or compound atoms are to be conceived as made up of a definite number of these simple atoms, arranged regularly in planes or cycles, so as to exercise polarity at their sides and angles. The solid state of bodies is that in which these compound atoms are attached to each other in permanent positions by their polar qualities, so that a slight increase of distance generates a powerful attractive force, and a slight decrease by pressure a still more powerful force of repulsion. The liquid state is that in which these compound atoms revolve on axes, so that their polarity and fixity of position disappear, and they are able to move amongst each other, while they still continue within the cohesive limit, or their attraction is increased by a slight increase of their mean distance. The gaseous state is when they pass by expansion beyond this limit, and the attraction being lessened by increase of distance, tend continually to separate more widely, and are only restrained within moderate distance by external pressure. The fourth, or igneous state, is that in which gaseous elasticity has ceased through further increase of distance, and the consequent absorption of *vis viva,* and in which, the solid atoms, having been separated into their planes, and even these, perhaps, into their simple atoms, these last are distributed almost uniformly by the repulsive force of the ether which they have separately condensed around them.

31. Resulting Definitions.

Monads are the self-repulsive particles of ether, diffused through all space.

Atoms are the dual particles of matter and ether combined inseparably, which constitute the first or ultimate elements of all ponderable substance. They combine in their mutual action the three different laws of general gravitation, cohesive affinity, and ethereal repulsion.

The *First Constant*, or neutral distance, is that at which two such atoms neither attract nor repel each other, or rather, attract only with the infinitesimal force of general gravitation.

Compound, chemical, or *plural atoms*, are the ultimate elements on which the constitution of every specific kind of gas, solid, or liquid depends, and must be conceived to consist of a definite number of simple atoms, in some simple or compound arrangement.

Sensible heat is the *vis viva* of the slight oscillations of solid or liquid atoms, or of the motions of the particles of gas amongst themselves.

Heat of fluidity is the *vis viva* of the motion of rotation in the particles of fluids.

Heat of vaporization is the *vis viva* absorbed in gases, by the separation of their elements to a wider distance, beyond their mean distance in their liquid form.

Heat of chemical combination is the *vis viva* developed by the new and closer relation of mutual distance, into which the components of compound atoms are brought by their union. Assuming the arrangements to be known before and after that union, its amount will be capable of exact calculation, the laws of affinity and repulsion being also known.

Light consists of vibrations or undulations, transverse to the axis of the rays, propagated through the free ether of space, and transmissible, in some cases, through the substance of solids and liquids either through interstices filled with free ether, or the combined ether of the particles themselves.

Electricity consists in the increase or decrease of the mean *vis viva* of ethereal repulsion between surfaces of material substances, arising from some change or special modification, of those surfaces, with reference to their charge of attached or combined ether.

CHAPTER IV.

ON THE IGNEOUS FORM OF MATTER.

32. *To explain the phenomena of the nuclei and tails of Comets.*

"There is beyond question," Sir J. Herschel remarks, "some profound secret and mystery of nature concerned in these phenomena. Perhaps it is not too much to hope that future observation, borrowing every aid from rational speculation, grounded on the progress of physical science, will ere long enable us to penetrate this mystery; and to decide whether it is really matter in the ordinary acceptation, which is projected from their heads with such extravagant velocity, and directed from the sun as its point of avoidance." He then remarks, on the tail of the comet of 1843, which was brandished unbroken in two hours through an angle of ninety degrees, and still reached to the earth's orbit; that "it is utterly incredible that it is one and the same material object, and that the notion of a negative shadow would best explain it; while still there are many other phenomena, as the issuing of the streamers from the nucleus, which link themselves just as irresistibly with our ordinary notions of matter."

In the Principia, Sir I. Newton argues that the tails of comets are rarefied smoke or vapour, which ascends from the sun, as smoke ascends in a chimney. But how it should ascend, unless the medium were much denser, instead of rarer than itself, or how it should acquire such a vast velocity, he does not attempt to explain. But he quotes the opinion of Kepler, that the rays of light might carry away the particles of matter along with them; which, after all, is probably the nearest approach to the truth which it is possible to make, while the constitution of matter and ether is unexplained. Let us now see whether light is not thrown on this subject, by tracing, on the present theory, the consequences of the igneous form of matter.

The mass of comets, it appears from many signs, is very small in proportion to their size. The matter they contain must then be highly rarefied. The idea of Newton, that "they are solid, compact, fixed, and durable, like the bodies of the planets," has been disproved by later and closer observation. Their density, it seems, is far less than that of air at the earth's surface. They must then, in their aphelia, be mainly in the igneous form, their particles widely separated, clothed with a full charge of ether, feebly united by gravitation, with a very slight amount of cohesive power.

When they return towards the sun, gravitation will ensure their near approach to a spherical form, and a process of gradual condensation. In those which are large or dense, this contraction will be more rapid and unequal, from the development of affinity also. As soon as this is called largely into play, it will cause intestine motion, and the *vis viva* produced will expand the less attracted parts, and retard their condensation. The comet will thus consist of slightly coherent nuclei, gaseous envelopes of each little nucleus, and a common igneous envelope of the whole; or else of gaseous spaces, and the igneous envelope alone. The uniform arrangement, and partial coherency of the envelope, will maintain the gaseous elasticity by a constant pressure. The longer the attraction of the parts has been undisturbed, the more will the nuclei enlarge, and the elastic force of the gaseous atmosphere will be more developed.

Such a body, moving uniformly through a sea of free ether, will assume by its resistance the form of a slightly prolate spheroid, and the nuclei, by the same resistance, will of course be found in the anterior part of the whole moving figure.

33. Besides the atmosphere of the sun, which revolves along with it, the space as far as the earth's orbit is more or less charged with matter in a very rare state, by which the zodiacal light is occasioned. This matter also must of necessity be in the igneous form. Each shell of it is drawn inward by the sun's attraction, and must produce a pressure on the parts within, and thus its density towards the sun must increase by a similar law to that of the gaseous atmosphere of the earth. When the comet approaches the sun, the solar attraction, the heating power of the solar rays, and the resistance of this igneous atmosphere, will all increase together, and the following results must ensue.

First, the increased velocity of the comet, the sun's attraction, and the inertia of the igneous atmosphere, will cause together an increased pressure on the igneous envelope. So far as this results from the motion, it will be in the line of the comet's path, but so far as it depends on the two other causes, in the line of the radius vector. The inner parts will be condensed under this pressure, with a great increase of the cohesive action, by which the inner strata of the igneous envelope will be condensed into gas, and some of the nuclei resolved into it also. Thus we shall have a nucleus or nuclei in the anterior, refracting light; a transparent gaseous portion, and an igneous envelope, lessened in amount, and pressing strongly upon the gaseous parts within.

Let us now compare the description which Sir J. Herschel has given (Astr. § 560). "That the luminous part of a comet is something in the nature of a smoke, fog, or cloud, suspended in a transparent atmosphere, is evident from a fact often noticed—that the portion of the tail which comes up and surrounds the head, is yet separated from it by an interval less luminous, as if sustained and kept off from contact by a transparent stratum. This and other facts appear to indicate that the structure of a comet must be that of a hollow envelope, of a parabolic form, enclosing near its vertex the nucleus or head. This will account for the apparent division of the tail into two lateral branches, the envelope being oblique to the line of sight at its borders, and therefore a greater depth of illuminated matter being there exposed to the eye." The nucleus, thus described, must be solid or liquid portions, like smoke or fog, not fully resolved into elastic, transparent vapour. The middle or transparent part must be in the state of gas, and the conic and parabolic envelope are the parts of the comet in the igneous form.

34. The formation of the tail is simply explained on the present theory.

Let a, r be the perihelion distance, and actual distance of the comet from the sun. Then its velocity in its orbit, supposed parabolic, is $\sqrt{\frac{2}{r}}$, compared with that of the earth, or 96300 miles per hour, and $26\frac{3}{4}$ per second, at the earth's distance. Again, its velocity towards the sun is $\frac{1}{r}\sqrt{2r-2a}$, compared with the earth's velocity of 68100 miles an hour. Thus, when the comet is at the earth's distance, and its perihelion distance less than one half, it must displace, every hour, a column of the sun's igneous atmosphere, more than 68000 miles in depth, and its head will move into a region of greater ethereal density. The front will be more compressed by this igneous atmosphere, and will contract in size. The hinder parts, repelled by the gaseous elasticity, less attracted by the sun, and less compressed by the external ether, will be left behind, less influenced by the comet's own attraction, and therefore must expand. The hinder part of the gaseous atmosphere, by this expansion, will reassume the igneous form. The increased pressure in front, by the motion which dips into the igneous atmosphere of the sun, must cause reaction in the neighbouring columns, as when a stone falls into still water, and this will sweep back the lightly adhering ether near the comet's head, and thrust the expanding hinder part into the line opposite to the sun. Hence a tail will be developed, lying between a line drawn from the sun to the head of the comet, and the line of the comet's motion; but having the former direction for its tangent, where it joins the head, because the ethereal repulsion must act in that line with the greatest energy. The velocity with which the matter of the tail is thus repelled at first will be less than that of light, which depends on the total elasticity of the ether, but will bear to it some definite proportion, determined by the relative number of the particles of matter, and those of free ether. Thus the immense velocity with which the tails are produced seems to be clearly and fully explained.

The approach to the sun is most rapid at 90° from perihelion, but the velocity in the orbit is greatest at the perihelion passage. Now the development of the tail, except so far as it depends on the internal structure, is determined by these two causes, and its maximum increase will naturally and usually be somewhere between them. Accordingly, Halley's comet made its perihelion passage Nov. 16, 1835, was 90° from it about 62 days earlier or Sept. 15, and its tail had the greatest apparent length Oct. 15, or exactly half way between these two extremes. At the first it had hardly begun to be formed, and at the perihelion it had disappeared. But in this case the latus rectum was greater than the diameter of the earth's orbit, and the velocity to the sun was decreasing when it entered the sphere of the earth. So far as the generation of the tail depends on the heating and exciting power of the sun, it will be greatest just before and after the perihelion; and accordingly, in other cases, the development of the tail is then the most striking. This was eminently true of the comet of 1843, which was so remarkable for the smallness of its perihelion distance.

35. The streamers from the head of Halley's comet, in 1835, are another remarkable phenomenon, which is simply explained by the present theory.

"On Oct. 2 the nucleus, which had been faint and small, was observed suddenly to have become much brighter, and to be throwing out a streamer or jet of light from the part turned towards the sun. This ejection, after ceasing awhile, was resumed on the 8th with much greater violence, and continued, with occasional intermittence, so long as the tail itself was visible. At one time the jet was single, and confined within narrow limits of divergence. At others, it presented a fan-shaped, or swallow-tailed form, like a gas flame from a flattened orifice; and at others, two, three, or even more jets were darted forth in different directions. The direction of the principal jet was observed to oscillate to and fro on either side of a line to the sun, in the manner of a compass needle, the change being conspicuous even from hour to hour. These jets, very bright at their point of emanation, faded rapidly away, and became diffused, as they expanded into the coma; at the same time curving backward, as streams of steam or smoke would do, if thrown out from narrow orifices in opposition to a powerful wind, against which they were unable to make way, and ultimately yielding to its force, so as to be drifted back, and confounded in a vaporous train, following the general direction of the current." The conclusions drawn by Sir J. Herschel are these: "That the matter of the nucleus is powerfully excited and dilated into a vaporous state by the action of the sun's rays, escaping at the points of least resistance. That this process takes place chiefly in the part turned towards the sun, the vapour escaping chiefly in that direction. That it is prevented from proceeding by some force directed from the sun, drifting it back, and carrying it out to vast distances beyond the nucleus. That this force acts unequally on the materials of the comet, the greater part remaining unvaporized, and a considerable part of the vapour produced remaining in the neighbourhood, to form the head and coma." To these remarks it may be added that the bulk of comets contracts greatly in approaching the sun, which M. Valz ascribes to increased ethereal pressure; but Sir J. Herschel rejects that view, as requiring a solid envelope, and refers it to vaporization, by which a great part is rendered invisible.

Now it has been shewn that, when the comet approaches the sun within the earth's sphere, the increasing density of the sun's unattached igneous atmosphere must press with greater force upon the comet's igneous envelope. Hence will follow a contraction of the anterior part, where the nucleus is contained, and increasing elasticity of the gaseous envelope, which will also increase in quantity, by the conversion both of igneous and diffused liquid portions into elastic vapour. Thus both the causes, suggested by M. Valz and Sir J. Herschel, and not one only, will be at work, and will conspire to produce a contraction of the nucleus, while the former alone will alter the apparent size of the igneous envelope. The objection of Sir J. Herschel to the solution of M. Valz, that the exterior of the comet would require to be like a skin or bag, impervious to the compression, is at once removed by a reference to the laws of the igneous form of matter, which really approaches to this very character. For matter in this state, though immensely rare, instead of being repulsive, must be slightly coherent, while it also transmits ethereal pressure.

When the igneous envelope has been thinned by this compression, like a soap-bubble by the resistance of the air, as well as by partial abrasion, and conversion into gas on its inner surface, while the contained gas has grown more elastic by solar heat, and increased molecular attractions, the pressure in some parts of its surface may overcome the resistance, and the pent-up gas will then escape through the orifice in the envelope, like steam from the boiler of an engine.

As soon as it escapes, which must be on the side towards the sun or the comet's way, it will expand rapidly by its own latent heat or gaseous repulsion, and one part will assume the igneous, and another the minutely liquid form, like steam when it cools, and will charge itself with ether from the surrounding medium. Being thus etherealized, and in the igneous form, it will be driven away from the sun by a double power; the reaction of its own impact on a highly elastic igneous medium, and the general tendency to diffuse itself in agreement with the laws of that medium. For the excess of matter in any column must cause an excess of ethereal density, which must then diffuse itself in the direction of least resistance, or in the direction away from the sun.

36. The formation of the tail, on receding from the sun, admits of a similar explanation.

The gaseous state of the comet will have reached its height, either at or soon after the perihelion passage, and also the pressure of the medium itself. When this pressure begins to diminish, as well as the sun's attraction, the matter of the comet will tend to expand in all directions. Towards the sun, however, it will be strongly restrained by the pressure, and hence it must actually expand most in the line of least resistance, or away from the sun. The effects, then, will have a close resemblance to those on the approach. As the comet recedes more and more, the pressure lessens on all sides, and the expansion will become greater, till the comet disappears from view through its gradual loss of illuminating power.

Thus all the main phenomena of cometary motion and development, which have excited so deep an interest of late years, find an easy and complete general solution, in the present theory, by a reference to the necessary properties of matter in its igneous form.

37. The separation of Biela's comet admits, in like manner, of an easy and simple explanation.

Let us suppose the nucleus, or denser part of the comet, to have two centres of maximum density, in the neighbourhood of which the cohesive force has begun to be most developed. The approach to the sun, increasing the temperature and the external pressure, will accelerate this action, and the pressure of the gaseous envelope will force the parts of the two condensing nuclei nearer together. The vis viva thus generated, will react on the gaseous portion, especially in the line which joins them, and by its reaction repel them further from each other. If they lie at right angles to the head of the comet, neither of them will be strongly repelled by the front pressure where it is a maximum, and they will swell out the sides of the coma. But the gaseous elasticity, in the line of their junction, being lessened by this lateral separation, will no longer neutralize the strong pressure of the external medium on the opposite part of the igneous envelope. It will thus be forced inward, like an army pierced by the irruption of a strong column in the centre, and a new lateral force will thus be created, which will complete the separation. But before the equilibrium of each part can be restored, it is quite conceivable that strong electrical relations may exist between the two separated portions, so as to explain those alternations of size and brightness in the two companion comets, which added to the singular character of the phenomenon.

All the explanations now offered grow by necessary consequence out of the fundamental assumptions; that, besides ponderable matter, there exists throughout space a self-repulsive ether; that the attraction of matter for ether follows a higher law than the inverse square, and probably as high as the inverse sixth, so that every atom of matter is combined inseparably with one of ether, and that the self-repulsion of ether follows a still higher law. From these data it will follow at once, that matter, besides a solid, liquid, and gaseous, must also be capable of assuming a fourth or igneous form; and that the properties of this form are such as to account for the singular variety of phenomena connected with the transmutations of comets on their approach to the sun, and their return from it again, which have caused so much natural wonder and perplexity to those astronomers by whom they have been observed, and of which the comet of this present year seems to furnish a new and striking instance.

CHAPTER V.

THE NATURE AND PROPERTIES OF LIGHT.

38. THE Undulatory Theory of Light, maintained by Huyghens, and since revived and extended by Young, Malus, Fresnel, and others, has now the general assent of men of science. By its help a large variety of complicated phenomena have received a clear explanation. Still, the vagueness of the usual conceptions of the nature of the luminous ether, and of its relations to common matter, have left many parts of the theory in a state of great obscurity, and offer large room for further scientific progress.

The present view of matter and ether will naturally include, as one of its immediate consequences, the whole theory of the undulations of light. At the same time, by supplying a distinct and well-defined conception of the laws of ether, in relation to common matter, it will, if a true interpretation of nature, provide a key to many outstanding difficulties, and solve many problems, which, without its aid, would be perplexing and obscure. The simplest order is to begin with the facts already explained, and to shew their connection with the theory, and then to pass on to others, which still need an explanation.

39. The postulates of the common theory have been clearly stated by Sir J. Herschel (*Enc. Metr.*), and they all flow naturally from the hypothesis of the present work.

(1) *An excessively rare and elastic ether pervades all space.*

This results at once from the assumed law of repulsive force in the ether. A finite number of such monads, confined to a given space, will arrange themselves so that the collective repulsion shall be a minimum, or very nearly at equal distances, though slightly condensed at the surface and edges, and the whole must have a density very nearly uniform. For if any portion were denser than the rest, it would expand by the excess of the repulsive force, till it reached, as nearly as possible, a uniform arrangement. If the density were solely due to the nearest particles, it would be strictly uniform. But since the total action of the central particles on those not adjoining is greater than for those near the surface, there will be a very slight difference in their density. For all sensible distances the distribution will be sensibly uniform, and the action like that of a continuous and homogeneous fluid.

(2) *It pervades all material bodies, and occupies the intervals between their molecules.*

This statement, on the present hypothesis, would be strictly true, if the mean distance of the compound atoms of bodies were much greater than the first constant, and the distance of the free ether in space. The surrounding ether would then fill all the interstices of the solid or fluid atoms. It seems to follow, however, from the conditions already examined, and the phenomena of chemistry, that the mean distance of the solid atoms is very little greater than that of the free ether in space. In this case very little free or unattached ether can commonly enter between the pores or interstices of the heavier solid bodies. But, on the other hand, the theory teaches that every atom of matter is combined inseparably with one of ether, and these retain unaltered their own self-repulsive power, though modified in its effects by its union with the matter, to which it is thus attached. Hence ethereal vibrations must of course be propagated through the whole extent of these bodies, though greatly modified by the presence of the matter, and the special atomic arrangement.

(3) *Either by passing between them, or by its extreme rarity, it offers no resistance to the motion of the earth, planets, and comets, appreciable by the most delicate observations.*

This statement has now to be modified by the discovery of the retardation of Encke's Comet, and some other similar phenomena. The calculated amount is such that, if the comet were to move with the earth's velocity in the earth's orbit, its motion would be extinguished by the resistance in about three millions of days. Supposing, now, its size to be that of the earth, and its mass 40000 times less, or density one-tenth that of air at the surface, then the same cause would extinguish the motion of the earth in 300 millions of years. But the earth's motion must probably be communicated to the ambient ether by frequent revolutions, nearly in the same time. Hence the relative velocity may be lessened ten or a hundred times, and the resistance in the duplicate proportion. Still, the fact of a sensible resistance in the case of some comets opens a wide field for curious speculation.

(4) *The molecules are capable of being set in motion by the particles of matter, and of communicating motion to the particles which are adjacent.*

This property results directly from the assumed laws. It must equally follow that this action is insensible at all sensible distances. In the illustrative case proposed, the action of ether on matter is less than that of an atom of matter on matter at one inch distance, and this latter force is insensible.

(5) *It is less elastic in refracting bodies.*

This conclusion is drawn from the smaller velocity of the waves of light in denser substances. And it results at once from the present hypothesis. For the velocity of restoration, after a disturbance, or the modulus of elasticity, must be diminished by the whole cohesive power of the matter combined with the ether, since the two forces from the same centres have opposite signs, and the repulsion of the ether centres is lessened by all the attractive power of the matter with which they are combined.

(6) *The frequency of the pulses, or number of impulses made on our nerves in a given time, determines the colour of the light, and the amplitude of the excursions, its brightness or intensity.*

This statement, it will probably be found, requires to be partially modified. The intensity of light seems to be more truly defined by the *vis viva* of the vibrating pulse which reaches or enters the eye, than by the amplitude of vibration in a single line of particles. From this larger definition it would equally follow that, with a single row of particles, a double excursion would imply a fourfold intensity. But the corrected view, besides its greater simplicity, removes a difficulty which would otherwise arise, from the difference between common light, and light elliptically polarized.

Again, the length of the wave determines the refrangibility, but it is still an open question, if not decided by this time in the negative, whether this property and the colour are inseparable. The origin of the sensation of colour seems to be physiological, even more than physical, and is not easy to explain, so as not to desert wholly the natural analogy with the musical notes, or waves of sound. With these slight reserves, all the main postulates in the common undulatory theory of light result simply from the view of matter and ether, and the laws of their mutual action, here proposed.

40. *To explain the Reflection and Refraction of Light.*

In Airy's *Tracts*, or Herschel's *Treatise*, the truth of the two fundamental laws, for reflection and refraction, is shewn to depend on two principles. First, that the velocity of a wave is constant in the same medium, and is diminished in one of greater density, in the ratio of the index of refraction. Secondly, that the face of the wave must be reached in the same time by parallel filaments or pencils of the luminous wave, since in other directions the vibrations are extinguished by their mutual interference.

The just claim to be made on any special hypothesis of the mode of action, is that it shall supply an adequate cause for the decrease of velocity in the denser medium, and one consistent with the great variety of refracting power in bodies of different composition.

Now the equilibrium of any solid or fluid, on the present view, must depend on a balance between the pressure of the external ether, and the attraction of the particles of matter joined with those of ether, on the one side, and the centrifugal force, or *vis viva* of atomic motion, and the repulsion of the combined ether particles, on the other. It is plain that the elasticity will thus be diminished, beyond that of free ether, by the counteracting force of all the attraction developed by the nearer approach of the matter to the neighbouring ether. On the other hand, it will be increased by the greater development of repulsive energy arising from the smaller distance of the combined ether particles from each other, compared with that of the free ether in space. The effect will be to diminish the elasticity of the recoil, but most of all for the short waves, of which the recoil is most rapid, since it is the increase of compression which alone can develop a larger ratio of repulsive or elastic power.

41. *To account for the Dispersion of Light.*

Sounds of different pitch are conveyed through air with the same velocity, and the notes of a piece of music, from any distance, reach the ear in their proper order. But the velocity of light is found to vary, in passing through gases, fluids and solids, with the length of the wave, since the violet are more refracted than the red rays. Sir J. Herschel remarks on this difficulty: "Neither the corpuscular, nor undulatory, nor any other system yet devised, will furnish that complete explanation of the phenomena of light which is desirable. Certain admissions must be made at every step, as to modes of mechanical action, when we are in total ignorance of the acting forces; and we are called on, when reasoning fails, occasionally for an exercise of faith."

Prof. Airy makes a similar admission of the difficulty which presses here on the theory, and offers a conjectural explanation: "Part of the velocity of sound depends on the altered elasticity of the air by the development of heat from its compression. If the heat required time for its development, the quantity of it would depend on the time the particles remained in nearly the same relative state, that is, on the time of vibration. If we suppose some cause, which is put in motion by the vibration of the particles, to affect in a similar manner the elasticity of the medium of light, and the degree of its development to depend on time, we shall have a sufficient explanation of the unequal refrangibility of the differently coloured rays."

Another solution has been proposed by Cauchy and others, from the hypothesis of finite intervals between the ether particles. The conclusion drawn from analysis is, that "a variation in the velocity of light is produced by a variation in the length of the wave, provided the interval between the molecules of ether bears a sensible ratio to the length of the undulation." This appears to be accepted by Dr Whewell, in the *History of the Inductive Sciences*, as a probable and adequate explanation.

Both of these solutions still involve serious difficulties. And first, it is not easy to see how the condition required by Cauchy's formulæ can be fulfilled. The length of a violet wave is one sixty thousandth of an inch. But Frauenhofer has ruled lines on glass, only double that distance apart, and Dr Wollaston has formed a platinum wire of the same diameter, while gold may be beaten to a thinness only one-fifth the length of a violet wave. Hence, on this ground alone, the distance of metallic atoms must be considerably less than the length of the shortest waves of light. But the previous inquiry would lead us to the conclusion that the distance both of material atoms, and of the monads of free ether, is much less than one billionth, and may probably not be much higher than one trillionth of an inch. Hence the disproportion between this distance, and the length of the waves of light, is in all probability far too great to admit of the solution Cauchy has proposed, and others have accepted, for the unequal refrangibility of different waves.

A second objection seems equally decisive. For on this view the dispersion should increase, as the medium departs more widely from continuity. But the ether must be most condensed in solid bodies, more rare in the atmosphere, and most of all *in vacuo*. Hence, by the hypothesis, it would seem to follow that the dispersion must be least in dense bodies, and greatest *in vacuo*, which is just the reverse of the real truth.

The other solution is open to an objection of hardly less weight. The time of vibration, in the case of sound, is immensely greater, and a more subtle element is present than the vibrating particles themselves; and yet there is no sensible difference in the velocity of long and short waves. Now both these differences would seem to make a variation more natural in the case of sound, where it does not exist, than in that of light, where it is so conspicuous. The first objection may be parried, by supposing the unknown cause, compared to the latent heat of the air, to need only a very short time for its development. But the second can only be removed, by supposing the dispersion to be due, not to a more subtile element, as the latent heat which accelerates sound, but to one less subtile than the vibrating medium, or in other words, to the ponderable matter. But the view then resolves itself into the early suggestion of Dr Young, that it depends on the reaction of the matter on the condensed ether, by which its elasticity may become a function of the time. This idea Sir J. Herschel seems to have dismissed rather too lightly. However vague such an explanation must be, in the absence of a definite law for the relation between matter and ether, it must seem highly probable on almost any reasonable view of that relation. On the present hypothesis, the truth of this early suggestion of that eminent philosopher will be fully confirmed.

42. Let us first consider, on this view, what will be the general condition of the ether, *in vacuo*, in gases, in liquids, and in transparent and opaque solids. *In vacuo*, it will be free, equally diffused, and intensely self-repulsive, the amount of ethereal pressure on the square inch being not much less than eighteen billions of pounds. In gases, it will clearly approach to its condition *in vacuo*, since only about one thousandth of the space is occupied by the material particles, and all the rest must be filled up with the ambient ether, slightly condensed by the cohesive affinity of the gaseous particles, while the simple component atoms of matter are also each combined with one of ether. In liquids, the proportion of free ether must be very much less. It results from the previous inquiry that the size of the compound material atoms can exceed, only in a slight degree, the intervals of the free ether in space. But since each atom of the fluid is in a state of revolution, and around its axis of greatest moment, each requires an atomic space equal to the cube of the mean distance of the particles or compound atoms, and this must exceed slightly their greatest length or breadth. The solid space, in the case of regular cubes, will be only $\frac{1}{9}\sqrt{3}$

or less than one-fifth of the whole, and still less for flat or oblong figures. All the rest of the space must be occupied by condensed ether, of which part may be attached to the fluid atoms, and revolve with them, and the rest fill up the interstices by the force of the immense ethereal pressure.

The state of the ether, in solids, and its relative amount, must plainly depend on the arrangement of the monads that compose the chemical atoms, and on the arrangement of those atoms amongst themselves. In other words, it will depend on their chemical composition, and also on the presence or absence, or on the peculiar kind, of crystallization. It is easy to conceive that, in the case of the more solid atoms, which have most cohesive force and the greatest density, there may be only single lines of attached ether, by which ethereal pulses can be directly conveyed across sensible distances. On the other hand, when the atoms have an angular or rhomboidal arrangement, there will as plainly be lines or paths of communication filled with ether by the ambient pressure, throughout the whole mass. The former will represent the class of opaque, the latter of transparent solids.

43. The velocity of a wave or pulse of ether, in travelling through a liquid or transparent solid, will plainly be affected by two causes. First, its path must wind round the material atoms, and since these, in their greatest length, nearly touch one another, there must in every case be a very sensible retardation, due to this cause alone. Thus, if we assume the atoms of water to be cubes, and the path of each ray so deflected as to move in circular arcs, of which the depth is to the chord as the side of a square to its diagonal, the path will be increased, or the

velocity lessened in the ratio of 1·3332 to one, which is very nearly the known refracting index.

Again, the undulations of light vary from 458 to 727 billions in a second. Supposing these to be transverse to the path of the ray, the view which results from the phenomena of polarization, then such excursion must be from the central path of least solid resistance towards the sides of the path, where it is shut in by the solid or fluid atoms. The ether will there be more condensed, in consequence of the cohesive attraction, and a motion through an equal space will therefore develop a greater resisting or repulsive force. And this difference will naturally increase, in proportion to the weight or inertia of the material atoms, though it must depend also on their arrangement.

It seems, then, to result at once from the hypothesis, that those vibrations which occupy the most time will also have the highest modulus of elasticity, and that this will be greatest for the longest waves. But a correct view of the chemical structure will clearly be requisite, before this variation can be theoretically traced in particular substances, so as to be compared with the results of experiment.

On this view the hypothesis of finite intervals is still virtually maintained, but in a modified form. It is not likely that the intervals of the ether atoms should have a finite and sensible proportion to the length of a violet wave, a distance almost sensible. But, in the present modified view, there is a finite ratio between the *excursion* of the ether particles, and the interval which separates adjacent material atoms. Each of these is probably nearly a billion times less than the length of a wave, and they may not be widely unequal to each other. For the linearity of the equation for the pulses of light implies that the duration is insensible, compared with the wave's length; while the previous comparison of known conditions leads to the inference that the mean distance of the material atoms cannot very greatly exceed that of the atoms of free ether.

44. *To explain the Polarization of Light.*

Polarization, in the Undulatory Theory, is accounted for by the transverse direction of the luminous vibrations. There are two directions, at right angles to the path of the ray, in either or both of which the particles may oscillate. Light is said to be polarized in any plane, when all the vibrations in that plane have been extinguished. When a ray travels horizontally, and is polarized in the vertical plane, the monads are conceived to move horizontally, at right angles to the line of the ray; and when polarized in the horizontal plane, the vibrations are held to be vertical. Very many remarkable phenomena have been thus explained. There are, however, some difficulties that have still to be removed, and Sir J. Herschel has stated them in these words.

"It may be objected that the molecules of the ether, if it be a fluid, cannot be supposed to be connected in strings or chains, but must exist independent of each other. But it is sufficient to admit such a lateral adhesion (we hesitate to call it viscosity) as may enable each molecule not only to push those which lie before it in the plane of its motion, but to drag along with it those on either side. It is true the properties we must attribute to the ether appear to belong to a solid rather than a fluid, and may be regarded as reviving the antiquated doctrine of a *plenum.* But if the phenomena can thereby be reduced to uniform and general principles, we see no reason why that, or any still wilder doctrine, may not be admitted; not indeed to all the privileges of a demonstrated fact, but to those of its representative or *locum tenens*, till the real truth shall be discovered."

The difficulty, from the implied resemblance of the ether to a solid rather than a fluid, is more apparent than real. The feature which distinguishes fluids from solids, is the want of polarity, resulting probably from the rotation of the atoms. But the monads of ether can have no such motion, and have therefore no true analogy with those of fluids. Their action on each other, under the assumed law, will have all the marks of strong polarity. If disposed in cubes, the repulsion along the side will be 64 times greater than along the diagonals, and 729 times greater than along the diameters of the cubes. The seeming difficulty, so far, is only a necessary result of the assumed theory.

The chief difficulty consists, not in the presence of strong polarity, like that in solids, but in the repulsive character of the force. For points united by a strong attraction have a close resemblance to an elastic cord; but repulsive forces tend to increase any actual divergence, and the restoration must depend on some external pressure. The transverse direction is here, too, a further difficulty. We cannot easily conceive them separated from direct vibrations still more powerful, since the direct must probably exceed the lateral force of restoration.

45. Let us first inquire what will be the structure of the ether, or the general law by which its particles will be arranged. Assuming the mean distance of the particles to be constant, they will be packed as densely as possible; or a given number in a given space will be so arranged,

that the mean distance of the nearest shall be a maximum. But the mean distance being the same, the density in a tetrahedral is to that in a cubical arrangement as 3 to 2 nearly. Hence the ether must assume a tetrahedral structure, and in the same plane will lie in equilateral triangles, or in radial lines at 60^0 of inclination.

Again, for small displacements, the direct force will vary as the displacement, and the lateral, as that displacement into the cosine of the inclination. Hence, in the actual arrangement, when a particle pushes others in a direct line, the forces in the two inclined directions will be one half on each side, or one half of the *vis viva* will be transmitted in the direct line, and one fourth in each of the two lateral lines. The motion, then, excited by a few particles cannot be transmitted to any sensible distance, since it would be reduced to one thousandth part after ten or twelve intervals.

Considering only the particles in the same plane, in a tetrahedral arrangement, three fourths of the *vis viva* of any impulse will be propagated in the direct line to the next monads, and one fourth in the two lateral directions at right angles. Again, supposing a circular disc of some size to be moved suddenly forward, it is plain that the whole *vis viva* will be communicated in the forward direction, except what escapes laterally at the periphery of the cylinder, or that which forms a reactive or backward wave, from the lateral recoil. The impulse will become weaker and weaker, with the distance travelled, even in the front, and will decrease rapidly in intensity, as the line diverges from that of the first disturbance. Every direct oscillation will produce one at right angles, and it seems to follow, from the fundamental law of arrangement; that both will be contemporaneous, or of equal elasticity, that the *vis viva* of the direct, will be double that of the transverse motion, and the last be equal in the direction of three axes at right angles to the path of the wave and at an angle of 120^0 with each other.

So long as a wave travels *in vacuo*, the ether will be symmetrically disposed, and the lateral vibrations which accompany the direct disturbance will have an equal movement in every plane which passes through the axis, or line of the wave. But when the ether is permanently condensed or compressed in certain lines, as in transparent solids, then the lateral disturbances, being at right angles to the path of the wave, will vary in their character, according to the internal structure. For the path of the wave, by the general laws of mechanics, must be in the line of least resistance, or the direction in which the ether is freest to move. The direction at right angles to this line of least resistance must be that of greatest confinement, or where the elasticity increases rapidly with the amount of displacement. Hence, while the path, being no longer direct, will cause a general retardation of all the waves, direct or transverse, the transverse waves, when longest, or the duration is greatest, will experience a relative acceleration above the direct waves, from this increased elasticity.

46. *To explain Circular Polarization, in contrast with Common Light.*

There is a difficulty, stated by Professor Airy at the close of his Tract, in explaining the distinction between common light and light elliptically polarized. "The most general kind of light that we can conceive," he there remarks, "is elliptically polarized, as the union of any number of vibrations in any directions, and following at any intervals, will produce it. Common light, then, must be elliptically polarized. The phenomena of interference compel us to suppose that common light consists of elliptic vibrations, many of which are exactly similar. Yet the difference of the phenomena shews that common light does not consist of an indefinite number of similar elliptic vibrations." The solution he proposes is, that in common light some thousands of waves are of the same ellipticity, after which there is a sudden and abrupt change to waves of a different species. The waves of the same kind must be numerous enough to produce all the coloured rings, but be compensated by those of a different kind in every single impression made upon the eye.

This notion, however, of a series of incessant and abrupt changes in all common light is very improbable, and hardly consistent with mechanical laws, which would require the change to be gradual, and not discontinuous. But the difficulty seems to arise from the assumption that a ray consists of one filament only, and that the intensity of the light is measured by the amplitude of the vibration alone. In that case an elliptic vibration would be the most general description of a wave. But if the intensity depends on the *vis viva*, or Σmv^2, for many related filaments, their vibrations may be in different planes around the common axis, and the difficulty ceases. Common light, on this view, may be called cylindrically polarized, each filament being polarized in one plane, but those planes radially disposed around the axis of propagation; while, in circular or elliptic polarization, each vibration is a circle or ellipse, and the particles form a helix or spiral.

Let us suppose an exciting cause, by which a series of

waves are transmitted from A towards B. After a short time there will be a balance between the supply of *vis viva* from A, and its dispersion laterally along the line. There will be a direct undulation along the axis, and secondary transverse vibrations at right angles to it, propagating themselves in the same line with equal velocity. All the displacements may be assumed to be infinitesimal, compared with the length of the wave. Every pulse of light will imply an axis of direct vibration, accompanied by transverse vibrations, collectively only one third in amount, but of the same velocity, performed in every plane which passes through the axis of onward motion.

The polarization of light will consist in the extinction of all the transverse vibrations, except those in one and the same plane. Two such waves, polarized in planes at right angles, and combined at the interval of one fourth of a wave, will produce circular polarization, or elliptical polarization, if the two waves thus combined are of unequal intensity.

47. *To explain the probable relation between Light and Radiant Heat.*

Light, by the received theory, consists wholly of transverse vibrations. But it seems to follow at once, from the nature of an elastic medium, that these must be joined with direct vibrations, even superior in force; because, in a tetrahedral arrangement, the whole lateral *vis viva* is only one third of the total amount. Any theory must be defective, which leaves this direct vibration unexplained.

Now it is plain that light and heat are usually joined together in the case of luminous vibration. Let us assume that all the direct vibrations are included under radiant heat, and the facts will be at once reconciled to the claims of mechanical law. Bodies usually require a high heat to become luminous, and the red rays, which answer to the longest waves, and have the greatest momentum, are the first which are visible. In other words, the direct vibrations must reach a certain degree of intensity, before the transverse ones, always feebler, have power to affect the nerves of the eye.

Heat itself, however, it has been proved latterly, admits of being polarized. We must assume, then, still further, that all pulses constitute radiant heat, in proportion to their *vis viva;* but that only transverse ones, and these within fixed limits of length or duration, are sensible to our eyes. Radiant heat will thus include both direct vibrations, and transverse ones beyond the spectrum, besides the luminous rays themselves. Now both of these last will plainly admit of being polarized.

In the experiments of Professor Forbes, the heat from different sources polarized through plates of mica, gave the following proportions. Brass, at 390°, 19 per cent.; mercury, at 270°, rather more; incandescent platinum, 54 per cent.; water, under 212°, 5 per cent. By reflexion, Argand lamp, 55 per cent.; brass, at 390°, 61 per cent.; platinum, 63 per cent.

These results agree well with the present explanation, if we assume that dense substances have a coercive power over the direct vibrations, increasing with the cohesive force; just as two polished surfaces may slide easily on each other, but offer great resistance, to pressure, or on direct separation. The direct vibrations, in water, would be 95 per cent.; in brass, 81; in mercury, about 78; while in platinum, the densest and most cohesive of metals, they are only 46 per cent. of the whole amount.

48. An explanation has been proposed, in Professor Kelland's Theory of Heat, on the hypothesis that the vibrations of heat also are transverse only. The formula for the polarization of Light when α is the analyzing angle, I the retardation, and λ the length of the wave, becomes

$$E = a^2 \left(\cos^2 \alpha - \cos 2\alpha \sin^2 \frac{\pi I}{\lambda} \right).$$

For the maximum we have $I = \lambda$, and the effect becomes $E = a^2 \cos^2 \alpha$. The assumption proposed is, that in the case of heat $I = \frac{1}{4}\lambda$, or the length of the wave four times greater than for light. But this is plainly an oversight. From the nature of the formula, λ must be the length of the wave for the polarized rays themselves, and $I = \lambda$ for the maximum in every case. On the proposed hypothesis, since $\frac{\pi I}{\lambda}$ would be 45^0, the intensity would be $\frac{1}{2} a^2$ in every position, and polarization must entirely cease.

On the other hand, the facts agree perfectly with the explanation now offered. For it will result that the direct are more powerful, or have greater *vis viva*, than the transverse vibrations in every case but one, that of platinum. And here there is a simple key to the exception, the coercive power of this densest and most coherent of metals taking effect mainly, if not exclusively, on the direct vibrations, which begin at right angles to its solid surface.

49. *To explain the origin of the dark lines in the solar spectrum.*

These are held, by general consent, to be one of the most interesting, but also one of the most perplexing, of optical phenomena. "The bands are constantly in the same parts of the spectrum, and preserve the same order and relations to each other, the same proportional width and obscurity, whenever and however they are examined; provided solar light be used, and the prisms are composed of the same material. A difference of material, though it causes no change in the number, order, or intensity of the bands, or their place referred to the colours, causes variations in their proportional distance. In the light of stars, in electric light, and that of flame, similar bands are observed, but they are differently disposed, and the spectrum of each star, and each flame, has a system of its own, characteristic of its light, which it preserves unchanged in all circumstances."

The simplest hypothesis, to account for this singular phenomenon, is that some change passes on the light in each case, when it emanates from the sun, star, or terrestrial flame, by which the answering parts of the spectrum are absorbed and extinguished. Now whatever be the source of light, it must be connected with material atoms, and among these some one or two atomic intervals of distances may be expected to prevail. Hence vibrations, in which the lengths of waves are the simplest multiples of these distances, will be favoured and maintained by the frequency of new pulses, while others will be extinguished, or have their force decreased. The simplest multiples will be those by powers of two, the next by those of three, and so on for the lowest primes. Let us now, beginning with 10000, write down the numbers which have only 2, 3, 5 for their factors, and we shall plainly have a series with irregular gaps, where primes, or composites of higher factors occur. If two such series, with incommensurable units, are superposed, the gaps will be diminished, but some spaces will still remain. In passing, also, through a dense medium, each ray may probably have its breadth extended from a mathematical line to a limit depending on the distance of the nearest atoms. Thus, if the length of the refracted wave were greater than 1000 and less than 1000 times the distance of the atoms, the vibration might be forced into coincidence with these extremes, and its spectrum spread over this interval. Every source of light, on this view, will produce dark spaces in the spectrum, not filled up by waves of particular lengths; and with different sources their position and number may be expected to vary, as they are found to do.

50. *To explain the relations between Light and Sound, with respect to the varieties of musical tones, and of coloured rays.*

The view has been prevalent of late, that solar light contains only three fundamental colours, red, yellow, and blue, which overlap each other, and may be separated by media of different absorbing power for different kinds of colour. Thus Mr Hunt remarks in his *Report to the British Association of Science on Solar Radiation.*

"On examining the spectrum of seven colours, we must at once perceive that these are composed of three—red, blue, and yellow. Examining it with coloured media, we can detect extensions of these primary colours, and fair evidence is afforded that these bands overlap each other. If we look at it through a cobalt blue glass, a colour unseen by the naked eye becomes visible, the extreme red ray. This is evidently the result of a mixture of blue and red. By throwing the spectrum on turmeric paper, a portion is seen beyond the violet, to which Sir J. Herschel gave the name of the lavender ray. Now if we examine all the conditions of these colours, we find that the yellow ray blends with the blue, and produces green; that the blue becomes more and more decided, passing to blackness in the indigo, but that the red reappears, and mixing with the blue, produces violet, and yellow blending with the violet produces the lavender ray. On the other side, yellow mixing with red produces orange, and then the red, growing in intensity and purity, again blends with blue, to produce the crimson or extreme red ray."

Reasons, however, of much weight have been offered for the opposite view, that the rays are simple, and not compound. For if the intensity of the light depends on the amplitude or momentum of the vibration, it is not easy to see on what the colour can depend, but on the length of the wave; just as the pitch of a note depends on the frequency of the aerial vibrations. But the length of the orange wave is just as definite as that of the red, yellow, or blue. And since all, when combined, produce only whiteness, any three tints, chosen wide apart, may be so combined as to produce any other tint whatever. In fact, Dr Young selected red, green, and violet for the three fundamental colours.

Again, if two vibrations, with a different length of wave, are combined together, the result will be doubly periodical, its two periods being the half sum and half difference of the two separate rates of vibration. But the period which depends on the half difference must answer to a simple wave beyond the limits of the spectrum. Hence that which depends on the half sum will alone be sensible to the eye; and this answers to a wave of the intermediate colour.

The affection of sight, also, in those who can distinguish only two colours, is a great difficulty in the way of the hypothesis that there are three only, fundamentally distinct, so as to produce the rest by their combination. For here the eye is sensible to all the rays, but wants the power to discriminate red from yellow rays.

The hypothesis, also, of three distinct spectra, which overlap each other, departs wholly from the strong natural analogy between light and sound, and replaces it by the foreign idea of material substance, and chemical composition. Since light and sound result alike from vibrations, it is natural to expect some close resemblance between the laws of colour and musical tone. The attempts to trace it, by comparing the breadths of coloured rays in the spectrum with musical intervals, or as some have done, with three chemical elements, have also a very arbitrary and empirical appearance, and are inconsistent with the fact of a variable dispersive power. But it does not follow that the analogy may not exist, and be traced out in a way which shall be free from all these objections.

51. *Musical Tones and Colours compared.*

Light and sound both result from periodical vibrations, one of air, the other of ether; and the general sensation is attended, in each case, with a vivid perception of musical pitch, or of colour. There is concord and discord in sounds, and a perception, hardly less vivid, of harmony or discordance in colour also. Some ears are insensible to minute differences of pitch or tone, and some eyes are equally insensible to many variations of tint. There are three fundamental notes in the octave; and there are also, in the view of most, three fundamental and primary colours.

The contrasts, however, are hardly less striking. The range of sounds includes six or seven octaves, while the whole spectrum, so far as sensible to the eye, is less than one. The musical scale has a series of definite notes, while intermediate sounds create a sense of discord. But the spectrum, except in the black lines, is one gradation of tint, and still gives an impression of singular beauty. The mixture of two colours produces an intermediate tint, but the striking of two notes together, a definite concord or discord. Coloured rays, again, seem of themselves to create a vivid sense of beauty; while in music it arises chiefly from the combination of sounds, and a definite relation between them, and a small deviation excites a sensation wholly opposite, or that of dissonance and discord.

These contrasts and resemblances may be explained in the following manner. The vibrations of light, being a billion times more rapid than those of sound, lie wholly beyond the reach of separate observation; whereas the slowest, in music, are only sixty in the second, or hardly one remove from being capable of direct enumeration. Hence definite recurrences of the vibrations, as in the musical third, where every fourth and fifth coincide, will be directly cognizable, and produce a sensible impression of agreement and harmony. In the case of light, this approach to distinct perception of the component vibrations must be impossible, and the contrast between concord and discord, arising from slight variations of pitch, destroying sensible periodicity of the two pulses, will disappear.

Again, the frequency of the vibrations, and the nature of the medium, render a more sensitive organ necessary to apprehend them. This might be expected to limit the range of its perceptive power; as the eye, the organ of sight, is much smaller and more delicate than the hand, the organ of touch. Accordingly, the spectrum is found to be within the range which answers to the musical octave.

Thirdly, this limited range of power, if assumed to be constant for the same eye, or even for different eyes, must create a natural standard, to which every ray may be instinctively referred. For every vibration, by the laws of mechanics, will be attended by its octave, either above or below. Hence the rays just beyond the extreme red and lavender will form natural limits to the sensibility of the optic nerve. Accompanying one another at the distance of an octave they may produce no sensation of colour, but exercise an exciting power, so as to form a secret, natural standard for the rest; a natural pitch for the luminous tones, which will be a contrast to the arbitrary pitch, consequent on the wider range of the musical scale.

This easy assumption being made, the analogy, in other respects, may be carried out without much difficulty. The simplest ratios of vibration to the limiting or fundamental tone will occasion a specific sensation, associated instinctively with each part of the surface from which such vibrations emanate, which will form a primary colour. Secondary ratios, in like manner, will form secondary colours. Those colours will be felt to be pure and vivid in which the wave length is nearest to the defining ratio, and those impure which lie between them.

52. The simplest ratios in music have two for their base, and are formed with the next primes 3 and 5. Thus $\frac{2}{1}$ is the octave, $\frac{3}{2}$ the fifth, $\frac{5}{4}$ the third, and $\frac{9}{8}$ the major tone. Beyond this limit the ear is not sensible of concord in the same series, and $\frac{16}{15}$, not $\frac{17}{16}$, is the chromatic semitone.

Now in Light, if we assume 424 billions for the vibrations per second of the fundamental wave, we have these relations.

9 : 8 = 477 billions = red ray exact; 5 : 4 = 530 billions (535 Hersch.) = yellow ray; 3 : 2 = 636 billions (644 = *b* intermediate to blue and indigo); 4 : 3 = 565 billions (*b* in the green); 5 : 3 = 707 billions (H. centre of violet).

Thus the three primary ratios, answering to the major tone, major third, and fifth in music, correspond with red, yellow, and blue, which by general consent are the three primary and simplest colours. The ratios next in simplicity, where three is the divisor, or the musical fourth and sixth, correspond with green and violet, which are the colours next in order of simplicity, so that Dr Young selected red, green, and violet for the three simplest. The next ratio is 6 : 5 = 509 billions, and corresponds exactly with the orange ray, and in music with the minor third.

The agreement, then, between the scales of light and sound is complete, when once we assume a ray beyond the red for a natural unit of comparison, or a fundamental pitch in the series of luminous waves.

Again, the waves beyond either limit will be octaves of those near the opposite limit. They may be rendered luminous by union with another ray, because the new mean, from the half sum of the lines of their vibrations will be within the limit of visibility. In this case they must seem nearly identical in their properties with the last visible rays at the other limit, just as a note and its octave sound together almost like a perfect unison.

53. It will result, from this view, that the rays of the spectrum are not, strictly speaking, compound; but that specific ratios in their vibrations, referred to a fixed limit

of visual power, give rise to the sensation of each primary colour; and the combination of two of these, because the period of the half sum prevails over that of the half difference in the compound period, will resemble the tint half way between them, but with some defect in purity and brightness of tone.

Again, the most usual variety of the imperfect vision, called colour blindness, is the incapacity to discern more than two fundamental colours, yellow and blue. This is easily accounted for on the present hypothesis. For the next ratio to the major tone is 17 : 16, and is insensible as a concord, or as a fundamental shade of colour. Now if this limitation in the power of the ear or eye were carried one step further, then the eye will appreciate the ratios 3 : 2 and 5 : 4, but not 9 : 8. Consequently only yellow and blue, to the exclusion of red, will be left for the fundamental colours, of which all the rest will appear to be varied combinations.

54. *Luminous and Non-luminous Matter.*

Another subject, on which theory needs to throw fresh light, is the source and nature of luminosity. It is often assumed that the sun and stars are composed of matter different in kind from the substances with which we are familiar. The present hypothesis would exclude all possibility of such an absolute contrast. We are thus led to the inference that the sun's luminosity is due, in some way or other, to its immense relative mass, and its central position.

The chief known sources of light are combustion or incandescence, phosphorescence, that is, a weaker and insensible kind of combustion, and solar radiation.

In combustion it is plain that the chemical atoms enter

into new and more intimate combinations. The effect must be that *vis viva* is largely increased; which may take the form of rapid oscillation, that is, high temperature of solid products; rotation of the new atoms, or the specific heat of fluidity; or rapid translation, where the products are gaseous. But in all cases alike there must be those ethereal vibrations, from which pulses both of light and of radiant heat usually arise.

Again, the whole pressure of an atmosphere on its lowest stratum equals the weight of the column, and hence the total pressure throughout the column will be one half the square of the mass into the mean force of gravity. But the pressure on each part must equal the repulsive force developed. This, again, in a gaseous medium, must depend on the centrifugal atomic motion, which varies as the square of the velocity into the mass of each particle. The same product must represent the *vis viva* of gaseous motion, on which the power to excite vibrations must be conceived to depend. Hence it seems probable that the collective power to disturb the surrounding ether will be measured, at least in some rude approximation, by the total pressure, or the square of the mass in each column, multiplied by one half the force of gravity at the mean height.

The mass of the solar atmosphere is unknown. Let us assume that it bears the same ratio to that of the earth as the two spheres bear to each other, reaching from their centres to the distance of equal attraction. Then $\mu\sqrt{\mu}$ will be the relative mass of the sun's atmosphere, μ^3 its square, $\frac{\mu^{\frac{3}{2}}}{r^2}$ the relative amount for the unit of surface, $\mu^4 r^{-6}$ the relative luminosity, or collective pressure for the unit of surface, and $\mu^4 r^{-4}$ the relative amount of light. On this hypothesis, the luminosity of the sun would be 12127½ millions, compared with that of the earth, and compared with Jupiter $1\frac{1}{9}$ millions. By this formula, the earth would have a native light, forty thousand times less than that of the sun which falls upon it, and Jupiter, on the contrary, a native light, equal to or greater than what it receives.

Now if Venus and Jupiter shone only by reflected light, the amount they transmit from a unit of surface should be as 50 to 1; and when she is in her maximum brilliance, and Jupiter in opposition, the disc of Jupiter is very little larger in visible size than hers. Assuming the crescent to be one fourth of her disc, her light on this view should be twelve times greater. The disparity, however, it seems certain, is considerably less. And hence there seems to be some direct presumption that Jupiter, the "lustrous star" of astrology, has really a native light, distinct from that which he receives from the sun. The auroras, again, are a native earth-light, and offer a further illustration of the same view. Luminosity, it is highly probable, is some function of the mass of each sun or planet, and of its velocity of motion, or the height of its atmosphere, though the exact function by which the two are related may not be easy to ascertain.

CHAPTER VI.

THE CHEMICAL ELEMENTS IN GENERAL.

55. THE Elements of the ancients were fire, air, earth, and water, answering to four general forms or conditions of matter, which is either solid, fluid, gaseous, or igneous. Modern science, however, applies the name to those kinds of homogeneous substance, which have not at present been resolved into each other, or into any simpler form. The total number of these is rather more than sixty, but one or two are doubtful, and several others of rare and limited occurrence. Hydrogen, oxygen, nitrogen, chlorine and fluorine, are gases; bromine and mercury are fluid at common temperatures; and the rest are solid, but the heat required for their fusion differs widely, and carbon, one of the most important, is infusible. The greater part, about fifty in number, have sufficient general resemblance to receive the name of metals. In their density and other properties, however, they differ widely. Potassium and sodium are lighter than water, while platinum and iridium are more than twenty-one times heavier. When binary compounds are analyzed by a galvanic current, the same

constituent in the same pair always goes to the same pole. Those which go to the positive pole are called electro-negative, when compared with those which go to the negative pole. Hence all the elements may be arranged in a series, from the most electro-negative to the most electro-positive, nearly as follows:

Oxygen, chlorine, iodine, fluorine, bromine, sulphur, nitrogen, phosphorus, selenium, arsenic; chromium, molybdenum, tungsten, boron, carbon, antimony, tellurium, tantalum, titanium, silicon; osmium, hydrogen, gold, iridium, rhodium, platinum, palladium, mercury, silver, copper; uranium, bismuth, tin, lead, cerium, cobalt, nickel, iron, cadmium, zinc; manganese, zirconium, aluminium, yttrium, glucinium, magnesium, calcium, strontium, barium, lithium; sodium, potassium.

Oxygen, chlorine, iodine, fluorine, bromine, which are highest in this list, are supporters of combustion. Sulphur, phosphorus, selenium, boron and carbon, are combustibles. Arsenic, chromium, molybdenum, tungsten, antimony, tellurium, tantalum, titanium, osmium, are acid-forming metals. Then follow neutral metals. Silicon, zirconium, aluminium, yttrium, glucinium, are bases of simple earths; magnesium, calcium, strontium, and barium, of alkaline earths; and lithium, sodium, potassium, of simple alkalies.

Some of these elements exist in very large quantities. Others, especially some of the late-discovered metals, are found only in very rare minerals, or in very minute proportions. The recent ones, and little known, are donarium, didymium, erbium, terbium, ilmenium?, lanthanium, niobium, pelopium, norium, ruthenium, thorium, and vanadium. The immense contrast in the width of their diffusion, and the small proportion of some which appears to exist, is a first presumption that the present elements, or at least the greater number of them, are really compound, and not simple and ultimate kinds of matter. For in this case, it is plain that some combinations would occur only under rare and exceptional circumstances, while others might be expected to be of frequent and ordinary occurrence.

56. Each of these chemical elements is found to have an equivalent, or atomic weight. In other words, a list of numbers can be framed, which represent the relative weights in which they combine, either directly, or in some simple multiple, however various the pairs or triplets which enter into combination. These numbers are naturally inferred to represent the relative weights of the chemical atoms themselves, or those smallest portions, which, by their union with each other, form chemical compounds. The ratio between hydrogen, the lightest, and gold, the heaviest, is 1 to 197. Out of sixty, however, there are only five or six of which the atomic weight is more than a hundred times that of hydrogen, and more than one half are less than 50. On the other hand, the lowest after hydrogen, or carbon, is six times heavier than that element. Thus the distribution is more nearly geometrical than arithmetical. No atomic weights occur between one and six, about one-half between six and thirty-six, and the highest is less than 216.

Mr Dalton, the discoverer of Chemical Equivalents, not only assumed that they represented the weights of the ultimate atoms of the elements, but that these were spheres, and formed chemical compounds by juxtaposition. He invented diagrams to represent the order in which they might be combined. But these arrangements had no dynamical character, and could therefore account for none of the properties, either of the simples or of their compounds. The conception, then, in this its first shape, has remained sterile and unproductive.

Another hypothesis assumes that the chemical atoms are centres of attractive and repulsive force, varying as the inverse square, but differing for each in its constant of force, and also in the distance at which the force changes suddenly from attraction to repulsion. On this view, however, the number of arbitrary constants introduced is at least twice as great as that of the chemical elements, and all explanation of the atomic weights themselves, and of remarkable relations which they exhibit to each other, is set aside. The properties of the elements, also, might then be expected to graduate into each other, when arranged by the order of their atomic weights. But an inspection of the lists will shew that substances, differing little in their atomic numbers, differ widely in their properties, as fluorine and calcium, or chlorine and potassium; and that others whose atoms are very unequal, have a strong resemblance, as sodium and potassium, molybdenum and tungsten, and others. No explanation of the nature of atoms can be correct which does not shew dynamical consequences resulting directly from their union, and then varying by some other rule than the atomic weight alone.

57. *Atomic weights, in many elements, are multiples of the hydrogen unit.*

When the doctrine of equivalents was first established, attempts were made, by Drs Prout and Thomson, to shew that all were multiples of hydrogen, and most of them even multiples. The later, and apparently more exact experiments of Berzelius and Turner tended, in some degree, to overthrow this hypothesis, and to establish broken or fractional values for many elements. Yet even their experiments, rightly considered, left the presumption in its favour very strong in the case of carbon, nitrogen, oxygen, and sulphur, some of the elements which are widest and most universal in their range. More recently, the analysis of Peligot and others has re-established integer values for chlorine, iodine, bromine, silver, and titanium. Even as the result, then, of experiment alone, the view seems now rendered probable, that all or most of the atomic weights are integers, if hydrogen, or else half-hydrogen, be taken for the unit. The following integer values, including many of the chief elements, may now be viewed as tolerably well ascertained:

Carbon 6, Oxygen 8, Boron 11, Magnesium 12? Nitrogen 14, Sulphur 16, Fluorine 19, Calcium 20, Sodium 23, Titanium 25, Iron 28, Phosphorus 31? Chlorine 36? Potassium 39, Molybdenum 46, Cadmium 56, Tin 58, Tellurium 64, Uranium 60, Arsenic 75, Tungsten 92, Iridium 99, Mercury 100, Silver 108, Iodine 127, Antimony 129, Bismuth 71 or 213, Gold 197.

It is plainly the low atomic numbers, which yield the most decisive evidence for or against the acceptance of integer values; and in the case of carbon, oxygen, nitrogen, sulphur, fluorine, calcium, iron, the reality of the relation stands out in clear and strong relief. In some others, the diversity of the earlier estimates has tended to interfere with the evidence for its truth.

Now if all the atomic weights were proved to be exact multiples of that of hydrogen, the natural inference must be that they are compounds, and that a certain number of hydrogen atoms, combined in some especial way, form the atom of every other element. Again, if many of them are multiples of hydrogen, but others fractional, the natural conclusion will be that hydrogen itself consists of two, three, or some small number of ultimate monads, of which last all other elements are composed. One of these alternatives seems to result plainly from the results of the most exact determinations of the atomic numbers, and either of them is in plain agreement with the present theory.

58. *Isomeric Elements.*

Several of the chemical elements are capable of assuming different forms, distinguishable from each other by strong chemical characters. Thus "the crystals of sulphur exhibit two distinct and incompatible forms, an octahedron with a rhombic base, the figure of native sulphur, which is assumed when sulphur separates from solution at common temperature; and a lengthened prism, having no relation to the preceding, when a mass of sulphur is melted, and after partial cooling, the crust is broken, and the fluid portion poured out." Carbon has wholly different qualities, as charcoal, and as diamond, and its crystals are of two or three wholly different forms. Phosphorus, again, as amorphous phosphorus, differs widely from its usual structure, is red, opaque, and insoluble in bisulphuret of carbon, and is one fifth denser than common phosphorus. Phosphoric acid, again, though composed of phosphorus and oxygen in the same proportions, has three distinct forms, which electricity disengages, without change, from the salts into which they respectively enter. Oxygen, also, exists under a second form, as ozone, in which it has a sulphurous smell, and enters more readily into composition with metals. All these facts, in the case of substances reputed simple, tend strongly to justify the suspicion that they are really composite, and that their constituent atoms may vary their manner of combination.

59. *Simple and Compound Elements.*

The Four Elements, Hydrogen, Oxygen, Carbon, and Nitrogen, form a remarkable contrast with all the rest by the multitude of distinct combinations which they form, and which constitute almost the exclusive subject of vegetable and animal chemistry. Some thousands of such compounds have been separated from natural products, or artificially formed, and the total number of atoms which enter into the compound atom, and fix its combining equivalent, is sometimes not less than a hundred. On the other hand, the other elements, in general, form only binary compounds, and after two binary stages, we reach the limit of their combining power. Of these four elements hydrogen is the lowest, and carbon is next to it, in its atomic weight, while oxygen and nitrogen are among those which come nearest to them in the smallness of their equivalents. The remarkable contrast between their powers of multiplied affinity, and the more restrictive capacity for union of the other elements, points naturally to the conclusion that all alike are real compounds, except possibly hydrogen alone, and that the simplicity of the compound structure in these lighter atoms makes them capable of a wider range of secondary composition.

60. *Properties of Water, a presumption for the composite character of the simpler elements.*

Water was taken to be a simple body, and one of the four elements, till the time when it was decomposed by Cavendish and other chemists. Since then, it has ranked as a binary compound of oxygen and hydrogen gases, though totally distinct in character from both of its constituents. But this view, though the natural result of discovery in its present stage, presents great difficulties to a thoughtful mind, when compared with the great facts of the universe. The quantity of water connected with our globe is probably four thousand times greater than all the free oxygen of the atmosphere. Its immense quantity has plainly made it one chief agent in the whole geological history of our planet. We seem to have distinct proof of its presence in large quantities, in the planet Mars, and, there is little doubt, in still larger quantities in the superior planets. Its properties, again, as an almost universal solvent, seem to give it the same rank, for simplicity, among liquids, as hydrogen, from its lightness, occupies among the gases. Its combining equivalent also is less than fifty out of the sixty elements, and indeed there are only four substances, now reckoned elements, which rank below it in this respect, and two of these are its own components. Again, it is now admitted by chemists, to have many of the relations of a metal of weak affinity for oxygen, with which it is made to combine by a laborious process. All these facts agree best with the idea, that water is really, in a certain sense, the simplest of the metals, that the other metals, of higher equivalents, are also compound, though our present methods have not decomposed them, and that the strong contrast between the electric affinities of oxygen and hydrogen, enabling us, in this case, to resolve the higher into two lower atoms, has accidentally degraded water, at present, from its true and natural place, which is higher in order of simplicity than that of the metals themselves.

61. *Related Atoms.*

The suspicion that our present elements are not simple, but compound, has gained strength of late in the minds of many chemical philosophers. Professor Faraday has expressed his conviction that it is highly probable, from the close resemblance among several metals, and the constant increase of the number discovered. Dumas, again, has pointed out several triads of elements, in which there is a gradation of properties, with atomic weights nearly equidistant. Such are chlorine 35?, a gas; bromine 80, a liquid; and iodine 125—127, a solid; and again, lithium = 7 nearly, sodium 23, and potassium 39; sulphur 16, selenium 40, and tellurium 64. Molybdenum has close resemblances to tungsten, and just half its atomic number. Iridium and platinum are very near akin, and their numbers are almost exactly the same. All these facts are best explained by the supposition that the elements are really compound, and that these have especial resemblance in their mode of composition.

62. *To explain generally the formation of Chemical Atoms.*

Every simple atom, on the present hypothesis, consists of two monads, one of matter, and one of ether, united inseparably into one. The matter of each attracts the ether of the other by a law, assumed to be the inverse sixth, or triplicate of the law of gravitation; and the ether repels the ether by a still higher law, assumed to be the duplicate of the last, or the inverse twelfth. These laws, from their nature, imply a neutral distance, at which the one repulsion balances the two attractions, and which must be supposed very small, not more, probably, than one trillionth of an inch. When two such atoms approach nearer than this distance, they will repel each other strongly, but when further apart, they will have a strong mutual attraction. This excess of attraction will be a maximum at the distance $\sqrt[6]{2}$, and after that will gradually but rapidly diminish. A monad of ether will be *in equilibrio* at a distance from a material atom, equal to this distance of greatest attraction for two material atoms.

Two, three, and four simple atoms, apart from foreign pressure or motion, will thus be *in equilibrio* in a right line, an equilateral triangle, or a tetrahedron, with the neutral distance for the side. With any greater number there must be a partial compression, until the excess of repulsion from this cause balances the amount of the attraction between the non-adjacent atoms. But the compression will be slight in point of distance, though great forces may be developed. For a particle at double the distance will only add $\frac{1}{64}$ to the forces already balanced, and a compression $\frac{1}{384}$ will suffice to develop an equal excess of repulsion. Compound atoms of this kind will have their inner monads within the neutral distance, with an excess of repulsion, or in a state of tension increasing towards the centre.

Let us now suppose a motion of rotation around a given axis. The particles will dispose themselves in a plane at right angles to that axis, and tend rapidly to assume equal distances, or the form of a regular polygon. Its radius, apart from the centrifugal force, would be such as to make the side of the polygon slightly less than the neutral distance. But the centrifugal force, if equal to the repulsion of the ether, will enlarge the side nearly as $\sqrt[6]{2} : 1$. With every transfer of *vis viva* to other atoms, or to the neighbouring ether, the cycle will contract; and enlarge with its increase, so as to find always a position of equilibrium. Such a motion of rotation evidently supplies the place of a central atom. In other words, revolving cycles find their equilibrium without a monad in the axis of rotation, but when at rest, an atom in the centre will often be needed for stable equilibrium.

63. A cycle of six monads, with a seventh in the centre, will be *in equilibrio* at the neutral distance, apart from a slight compression due to the remoter monads. One of seven monads, with an eighth in the centre, will be *in equilibrio*, when the side is less and the radius greater than the neutral distance, and so also a cycle of more monads than eight. The monads will be attracted to the centre, and repulsive to each other, and any disturbance, which throws them out of the plane, will produce a new and more compact arrangement. On the other hand, if one monad be in the centre, and three, four or five in the circumference, the action will be reversed. The tangential action will be attractive, the central repulsive, and the tendency of disturbance will be to force out the centre atom to some distance from the plane. But a movement of rotation will alter all these relations. It will first transfer the excluded central monad to the plane of the cycle, and if increased further, from the centre to the circumference, so that the moment of rotation may be a maximum.

64. Compounds of several monads may combine in several different ways. If both solid, or composed of adherent monads, combined at nearly their minimum distance, they must unite by simple apposition, so that their centres may be as near as possible. Flat cycles might thus be joined face to face, and prisms and cylinders,

side by side, or end to end. If they are revolving, so as not to change their own shape, they will repel each other at their equators, and tend to unite, either pole towards equator, or opposite poles to each other, on account of the repulsion of the ether where the two approach nearest. But if one of them is composed of one, two, or more cycles, revolving round an axis, they may combine radially, so that the hollow cycle shall encompass the other compound atom. For such a combination it is plainly required that the *vis viva* of rotation or atomic latent heat shall be greater for the compound atom which excludes the ether after their union.

A further variety of combinations may result from the action of the free or uncombined ether. The neutral distance, it follows from the theory, is less, but not greatly less, than the mean distance of the compound material atoms from each other. It has been shewn to be probable that the mean distance of the free ether monads in space is of the same order, greater than the neutral distance, but less or not much greater than the distance of the compound atoms. Hence the ether which fills up the interstices of solids and fluids, must bear the relation to the separate atoms, not of an atmosphere to a planet, but rather of planets to a sun, or be relatively discrete, and therefore strongly polar. Separate ether monads will combine with the compound atoms at their periphery, and deport themselves like material monads, except that their neutral distance is slightly larger. Revolving cycles may thus be formed, of rather more than double the first radius, by the mixture of alternate ether monads, and will expand or contract by a similar law, as the rotation is increased or diminished.

65. The pressure of the free ether on the surface of all solids and liquids must produce a further effect in modifying the relations of the atoms to each other. In transparent bodies, it is plain that the vibrations of ether travel with comparative ease and freedom, and hence the ethereal pressure cannot possibly exercise any strong influence in rendering the structure more compact. But the reverse must be true in opaque bodies, where the opacity is one essential feature, as in the metals. The immense ethereal pressure must, in this case, tend to force the atoms more closely together, and thus to produce a tension, and an increased repulsion, at the periphery of those atoms, when they are *in equilibrio*. This will naturally increase the permanence of the groups, which compose the compound atoms, and thus remove the difficulty which might else lie against the present view of their formation, that the different elements must then be capable of easy conversion into each other, which would contradict all our present experience.

66. The permanence of the compound atoms will be further secured by their *vis viva*, or atomic motion.

Let the density of the earth be 5·5, and the atom of water 9, or hydrogen the unit. Then the gravitating force of one monad on another, at one inch distance, or the velocity in inches per second, will have $3n+7{\cdot}2267$ for its negative logarithm. Suppose the cohesive force about six times less, or its neg. log. $=3n+8$, the neutral distance 10^{-18} or one trillionth, and the mean distance of the water atoms about five times greater. Then $3n+8=60$ is neg. log. of cohesion at distance one inch, and $108-60=48$, pos. log. at neutral distance. But in circular motion $v^2=fr$, $\therefore$ $\log v=15$, or the velocity of rotation, to make the centrifugal force equal to the attraction or repulsion at the neutral distance, must be 1000 billions of inches, or 160 quintillions of revolutions per second. In revolving cycles with no central atom, the *vis viva* capable of being absorbed is plainly still greater, since the centrifugal force is balanced only by a very oblique attraction of the nearest atoms. Two such revolving atoms will plainly have an intense polarity. They will repel each other by their equators, and will be able to combine only at the poles, by parting with *vis viva* on one side, and absorbing it on the other. In this way two distinct atoms may combine radially, and *vis viva* will be produced by their union.

CHAPTER VII.

THE FOUR SIMPLEST ELEMENTS.

67. "The doctrine of the alchemists on the transmutation of metals," Professor Faraday observes, "is no longer opposed to the analogies of science, but only in some stages beyond their present development." The same idea has been carried still further by Professor Low, in his Inquiry into the nature of Chemical Elements. He concludes that they are all, probably, compounds of hydrogen, carbon, and oxygen, or possibly of hydrogen and carbon alone. But no approach has been made to any such view of their composition, if they are really compound, as will account for the immense variety of their distinct properties. An attempt will now be made, in the case of the simpler elements, to bridge over this gulf, and to make at least a conjectural approach to a complete theory, which will supply a real explanation, by the laws of mechanics, for the chief distinctive properties of these elements.

68. *To explain the disproportion between the atomic weight of Hydrogen, and the other elements.*

There are about 65 known elements, and sixty of them do not exceed 100 in the hydrogen scale of atomic weights.

The mean interval, then, is $1\frac{2}{3}$. But the lowest equivalent, after hydrogen, is carbon = 6, or the interval is three times the mean amount. In point of ratio the interval is much greater, being one third of the whole proportion between the lowest and highest equivalents. There must surely be some reason for this singular break which meets us at the entrance of the atomic scale.

Let us now assume that hydrogen, by far the lightest substance known, answers to the simple monad, or dual atom of matter and ether combined. Three is the smallest number which can form a closed figure in one plane, and two is the smallest number of planes which can combine to make a solid figure. There is thus a sixfold interval between the monad, and the simplest cyclical atom, which has a solid character. This is precisely the ratio between the equivalent of hydrogen, the lightest body known, and carbon, the lowest in atomic weight of the solid elements.

The same assumption accounts at once for the fact, now well ascertained, that many of the atomic weights are precise multiples of that of hydrogen. The cases, which appear to give a fractional value, may be reconciled with it in three ways—by the supposition of a faulty experiment, or of a double atomic weight; or lastly, by the conversion of some small portion of the element into hydrogen in the course of the experiment.

69. *Direct evidence for the greater simplicity of the Hydrogen Element.*

If Hydrogen be the simple monad, then every other element might be resolved into hydrogen by a sufficiently powerful analysis. Where the outmost monads of an atom are not compacted closely a partial resolution may be possible even by forces now at our disposal. Such a change, affecting a very small portion of the substance, might easily escape notice, and be referred to some other cause, before the likelihood of its occurrence had been shewn on other grounds.

Now Sir H. Davy "observed the production of a small quantity of water during the combustion of the purest charcoal, indicating a trace of hydrogen." "But its quantity," Dr Turner adds, "is so small that it cannot be regarded as a necessary constituent. It proves only that a trace of hydrogen is retained with such force, that it cannot be expelled by the temperature of ignition."

This explanation is not at all satisfactory. There are many chemical affinities which yield to the heat of ignition. How, then, should a union merely mechanical, and therefore much less intimate in its nature, hold out against it? It seems more reasonable to allow that a small portion of the carbon is really decomposed, since a trace of water appears constantly in the product.

"Sulphur, like charcoal, retains a portion of hydrogen so obstinately, that it cannot be wholly freed from it, either by sublimation or fusion. Davy detected its presence by exposing sulphur to the strong heat of a powerful battery, when hydrosulphuric gas was disengaged. The hydrogen from its minute quantity, can only be regarded in the light of an accidental impurity, and in no wise essential to the nature of sulphur." Here, again, an opposite explanation seems far simpler, that, as Professor Low remarks, "a portion of the sulphur itself was decomposed."

A third fact of the same kind was noticed by Dr Prout, and confirmed by Dr Turner. "Chloride of silver, however white and well washed, gives out little muriatic acid at the moment of fusion. When it has been well dried at 300° and then fused, the loss is not appreciable in weight, though still sufficient to redden delicate litmus paper." This result might arise from a slight retention of moisture, even at that high temperature; but the analogy of the two previous facts adds weight to the presumption, that a few atoms of chlorine or calcium may be decomposed, and furnish hydrogen for the hydrochloric acid.

Again, the whole recent progress of chemistry has tended to replace oxygen by hydrogen, as the main agent in a large class of chemical compositions, and to disclose the immense variety of the proportions in which, along with carbon and oxygen, it builds up all vegetable and animal substance. These facts seem to point to the conclusion that it represents matter in its simplest and least composite form; while carbon, oxygen, and nitrogen come next in order of simplicity. And the atomic weights of all these are confessedly simple multiples of hydrogen.

70. *To deduce probably some leading properties of Carbon, Oxygen, and Nitrogen from their atomic numbers.*

These three elements, along with hydrogen, build up the ocean, the atmosphere, and the whole animal and vegetable world. They stand out thus in relief from all the rest by the immense variety of their combinations. Oxygen and nitrogen compose the air; oxygen and hydrogen, the water of the ocean, seas, and rivers; carbon, hydrogen, and oxygen, nearly all vegetable substance; while all the four commonly enter alike into the products of animal life.

The equivalents of these elements are expressed exactly by the integers 6, 8, 14. The densities, however, of nitrogen and oxygen gases are as 7 to 8; and the relation of carbon to nitrogen in cyanogen, their chief compound, is 6 to 7. The researches of Gerhardt and others in organic chemistry have also led them to prefer double values, 12 and 16, for both carbon and oxygen. The ratio of hydrogen and oxygen is 1 to 16 in the density of the gases, but 1 to 8 in water, their chief and only natural compound. Hence we infer that the relation of these elements to each other and to hydrogen may be expressed ambiguously by the three numbers 6, 7, 8, or by their doubles 12, 14, and 16.

Each of these numbers depends on one of the three first primes 2, 3, 7. The simplest prime answers to oxygen, the most universal and active element; the next to carbon, and the third to nitrogen; which is just the order of their abundance and activity in combination.

71. Carbon, again, is the chief base, and oxygen the chief sustainer of combustion. Nitrogen has a kind of neutral character, and is marked chiefly by negative properties. "Electricity," Faraday observes, "determines no liberation of it from its compounds, which seems to point to the idea that it is not itself a simple substance. Its most distinctive feature is a backwardness to combine directly with other simples, and the weaker affinity of the elements in its chief compounds."

Let us now conceive 6, 7, and 8 particles disposed in one plane, and observe the differences that will follow. Their permanence in that plane implies and requires a moment of rotation; and conversely, a rotation sufficiently rapid, will make them form a single cycle or polygon in one plane. The side of a hexagon is equal to the radius. Hence, if each monad were acted on by the two nearest only, the hexagon of equilibrium would have the neutral distance for its radius. The action of the three further atoms will produce a slight compression. A single atom, approaching such a hexagon in the line of the axis would tend to increase the compression still further, and could not therefore find a place of rest in the centre of the plane, but might adhere to it in the axis at some little distance, until a stronger rotation increased the radius of the hexagon, so as to exceed the neutral distance.

Two single atoms might attach themselves to such a hexagon on either side, even when at rest, and will tend to increase its stability; since, when neutral to each other, they will be strongly attractive to the cycle itself. A single atom, also, might rest on the axis between two such cycles, and tend by its action to draw them nearer together. A similar arrangement is clearly possible in the case of several such cycles. Now it is well known that there are a series of hydro-carbons, the basis of many compounds, in which the equivalents of hydrogen are either one more, or one less, than those of carbon; as Formule C_2H, Methule C_2H_3, Acetule C_4H_3, Ethule C_4H_5, Mesitule C_6H_5, Glycerule C_6H_7, Amyl $C_{10}H_{11}$, and others. There is here, at least, a *primâ facie* correspondence with the natural properties of a hexagonal cycle.

"Light Carburetted Hydrogen is not decomposed by electricity, nor by being passed through red hot tubes, unless the heat is very intense; when some of the gas is decomposed, and yields two volumes of hydrogen and a deposit of carbon." Such a result seems to flow naturally from the hypothesis. Its formula CH_2 seems to answer to the case of a hexagonal cycle, with an atom axially joined on each side. A high temperature answers to a more rapid rotation, by which the radial atoms would recede further from the centre. When the centrifugal force exceeds the attraction of the non-adjacent atoms, the radius will exceed the neutral distance, and the hydrogen atoms will have nothing to retain them on opposite sides of the centre, but can slide along the axis with the least disturbance. They may thus be driven out and separated from the hexagon, while the *vis viva*, being absorbed or sensibly lessened by this separation, may change the hexagon into two triplets of atoms, the first and simplest element of a solid structure.

72. A cycle of eight will have relations entirely different. The radius of an octagon, when the side is one, will be 1·307, and the reduced radius, after the compression resulting from the assumed laws, will be 1·267 nearly. Hence another monad might find a place in its centre, but not in fixed or stable equilibrium. Before another atom on the axis can enter into fixed and permanent union, the eight-fold cycle must be decomposed or divided into two quaternions, when the nine will assume their most compact form in the shape of a cube, with eight atoms at the corners, and one in the centre. Such a particle has plainly the elements of a solid structure; but since its three axes are all equal, and its moment of rotation is little more than one third of its former value in the first arrangement, it will be clearly most favourable for the assumption of a fluid form.

Of all substances now reckoned compound, water is by far the simplest, the most widely diffused, and the most extensive in its powers of combination. Water and fire, by general consent, have been the two main instruments in determining the present structure of our globe. The mere fact that it is now held to be a compound, while a score of metals, found here and there in minute particles, and separated with much labour from complex minerals, are viewed as simple, is a clear sign that chemistry is only in a temporary and provisional stage.

The atomic number of water is 9, and is next in order to 6, 7, 8, those which characterize carbon, nitrogen, and oxygen, in its relative simplicity. Measured by this test, water is really simpler than all, except these three or four primary elements, and a very few of their compounds, whether capable or incapable of analysis by processes actually known. This number is the sum of 1 and 8, the numbers of hydrogen and oxygen, but as the square of 3, or 3.3, it is intermediate in character between 2.3 the number of carbon, and 2.4 or 4.4 the two values of oxygen, when these are similarly expressed. Carbon is only known as a solid, oxygen only as a gas, and water is intermediate in its properties, being the type of a fluid.

A peculiar amount of heat is generated in the formation of water. The oxy-hydrogen blowpipe thus becomes one of the most powerful of chemical instruments. The view of this combination now offered supplies a clear and simple explanation. The contrast will here be greater, than in other cases of combustion, between the latent *vis viva* of the simples and of the resulting compound. It is not simply that the oxygen and hydrogen combine, but that the oxygen monads, which combine with other elements in an eight-fold cycle, in this case alone are brought one step further, and become really four-fold. We should infer that, for the same amount of matter, the heat evolved in the combustion of hydrogen must be greater than for any other element. On the other hand, for the same amount of oxygen more *vis viva* may possibly be set free in some other cases, because the oxygen monads approximate to a larger number of monads in the other element. The experiments of Despretz give the following amounts of heat for the combustion with equal quantities of oxygen. Hydrogen 2578; carbon 2967; iron 5325. Phosphorus, zinc, tin, nearly as iron. The amount increases with the atomic weight of the combustible, but when this reaches a moderate limit, becomes nearly constant. The monads of the larger atoms, with which those of the oxygen are most nearly in contact, on which nearness the evolution of *vis viva* depends, cannot exceed twice or three times its own number. But the amount of matter being the same in the product, the values become—hydrogen 286½, carbon 269, iron 184, zinc 133, tin 184; where hydrogen occupies the first place. Again, Dalton found that hydrogen, in burning, raised its own volume of water five degrees, and light carburetted hydrogen, eighteen degrees. The latter combines one atom of carbon and two of hydrogen. This gives the ratio 5 to 8 for the heat from one atom of hydrogen to one of carbon, or about two thirds, atom for atom, and three times as much for an equal weight.

73. The density of oxygen gas is sixteen times greater than that of hydrogen. Hence, on the present hypothesis, we must assume that the gas consists of two cycles of eight, or one of sixteen monads. The equivalent from gaseous density being thus double that inferred from the oxides, has caused a great schism among chemists, whether eight, or its double, sixteen, is to be preferred.

A second important fact has been established of late years, that oxygen is allotropic, and exists in two different forms. By the action of electricity, the slow combustion of phosphorus, or contact with the heated vapour of ether, it becomes ozone, acquires a sulphurous smell, the power of oxidizing silver and other distinctive properties. By passing it through red hot tubes, it is reconverted into common oxygen. At the same time more recent researches have seemed to imply the existence of a teroxide of hydrogen with properties very similar, and hardly distinguishable, from those of ozone.

The two facts, of the existence of ozone, and of the double equivalent of oxygen gas, seem to throw light upon each other. A cycle of sixteen, or two cycles of eight, must be convertible into each other by increase or diminution of the centrifugal force. Now ozone is produced from common oxygen by the electric discharge, and resolved into it again by a red heat. Increase of temperature implies an increase of *vis viva* and therefore a tendency of the atoms to assume the form in which the movement of rotation is the greatest, or that of a single cycle in the same plane. But the transmission of an electric shock or spark seems to imply a contrary relation of the atoms, in which they turn edgewise to each other. This implies a temporary rotation on some diameter, and a force acting for an instant at right angles to the principal plane. The tendency of such a movement must be to change the one cycle into two of half the size, or it may be, two cycles of sixteen of alternate matter and ether monads. Hence it seems probable that oxygen gas may consist of single cycles of sixteen material monads, and ozone of two cycles of eight, or else two of sixteen, half material, and half of uncombined ether. On this view ozone, which resembles sulphur in its smell, is also a step nearer to a solid structure than gaseous oxygen. Also the atomic weight of sulphur is 16, just double that of oxygen, so that it is no improbable conjecture that it may consist of two particles of ozonized oxygen united into one.

74. Oxygen is the great supporter of combustion. But the cause of the heat thus produced is left wholly unexplained by the theory which regards it as a distinct and ultimate element. Dr Turner observes, after noticing the defect of Lavoisier's explanation.—"A new theory is therefore required to account for the chemical production of heat; but it is easier to perceive the fallacies of one doctrine than to substitute another which shall be faultless."

The present hypothesis, I believe, provides a full and complete explanation. Heat is simply atomic or molecular *vis viva.* Sensible heat depends on the oscillations of the solid atoms, transferred through the repulsion of their constituent or adjacent ether to neighbouring atoms, and producing vibrations which can radiate freely through air or *in vacuo,* like the waves of light. Heat of fluidity consists in the *vis viva* of each atom in revolving on its own axis of greatest moment, whereby the polarity of neighbouring atoms is weakened or destroyed. Heat of vaporization consists in the *vis viva* spent or absorbed in removing the chemical atoms to a greater mean distance beyond the limit of maximum cohesive power, so that the centrifugal force is balanced only by some external pressure. But besides these varieties of heat, depending on the solid, fluid, or gaseous structure of the whole body, there will be another variety, an atomic heat, depending on the essential structure of the compound atoms themselves. It will be least where their monads are most compact, and greatest when they are at greater intervals from each other. The laws of force being given, its amount may be easily calculated for every variety of arrangement of the component monads. Thus with the law of the inverse fourth, if the unit of *vis viva* be the velocity of one pair of monads acquired in falling from infinity to the neutral distance, the value for an atom of 9 monads in three triplets would be 16 nearly, and that for one cycle of 16, 100, leaving an excess of 68 for the latent *vis viva* which would become sensible heat, or molecular motion, when the cycle of sixteen passes by combination into the other form.

75. Another remarkable feature of oxygen is its wide diffusion and great combining power.

Fluorine is the only element, of which some compound with oxygen is not known. Now its atomic number is numerically of the simplest kind, being formed by successive doubling only. The values, also, deduced from its solid compounds and water, and from its gaseous density, have this very same ratio to each other. This singular feature occurs in no other gas but the vapours of phosphorus and arsenic only. Again, there are thirteen or fourteen metals with which it forms a sesquioxide. Hence a suspicion may naturally arise, that a half atom here enters directly into combination. In this case the three numbers 4, 8, 16, formed by successive doubling, would all belong to oxygen in these allotropic forms, and serve to account for its immense combining power.

76. *Allotropic Forms of Carbon.*

Carbon, it has been discovered, admits of four allotropic forms. The first is the diamond, sp. gr. = 3·52 and octahedral. The second graphite, sp. gr. = 2·17, and its form the hexagonal prism. The third coke, sp. gr. = 1·89, and its primary form a cube. The fourth anthracite and charcoal, sp. gr. = 1·773, and its form square prismatic, the axes nearly as 5, 5 and 3. In the last form it is softer than coke, and a worse conductor of electricity. By a bright red or white heat it is changed to coke in its other characters, but not in its specific gravity or crystalline structure. What light is thrown on these striking and characteristic facts by the present theory?

First, the densest and most compact form in which six monads can be disposed is the regular octahedron. Now six is the atomic number of carbon, the diamond its densest form, and an octahedral crystal. On this view, also, the peculiar hardness and cutting power of the diamond are fully explained. The apex of every atom so formed must form a molecular wedge, to penetrate any surface to which it is applied. The shape of the ultimate octahedron must also reproduce itself, evidently, in the cleavage of the whole crystal.

The second form, in graphite, has for its type the hexagonal prism. Now if two cycles of six be superposed, we have clearly the first element of this very form. The number of monads in such an atom will be double those of the octahedron, but the area of the circumscribing sphere is $2\sqrt{2}$ times greater, and the actual area, when the two cycles are in their stable position, or $\frac{1}{2}\sqrt{3}$ their vertical distance, as $\frac{1}{3}\sqrt{2} : 2\frac{1}{4}$, or 1 to 4·772. Hence the density must naturally be less in the hexagonal structure, in a ratio between the limits 1·414 and 2·386, but from the crystalline and diaphanous structure of the diamond, much nearer the former limit. The experimental ratio is 1·62 to 1.

77. The two other allotropic forms, in coke, and in anthracite and charcoal, are the cubic and square prismatic, the latter having one axis about three fifths of the others.

The latter is converted into coke at a white heat, retaining otherwise its former structure.

Now if we suppose the two hexagons to be superposed exactly on each other, the double hexagons arranged symmetrically at right angles in the same plane, and the second tier also at right angles, but alternately in position, the height of each tier will be twice the neutral distance, that is, once between the two hexagons, and a half distance on each side. But the diameter, in each of the two horizontal axes, will be twice the neutral distance also. Hence it seems plain that the result must be a structure of which the type is the cube, or the three axes all equal.

Again, let us suppose a similar arrangement, with the one difference that the two hexagonal cycles in the same atom are alternately arranged. The vertical axis will now be $\sqrt{3}$ instead of 2, or in the ratio of 6 to 7. But the external attraction of the neighbouring atoms will tend doubly to enlarge this proportion, and may render the third axis unequal in a still greater degree. On the other hand, the application of heat, increasing the lateral repulsion, will tend to restore the atoms to the more vertical and less compressed position.

78. *Properties of Nitrogen.*

Let us inquire what will be the characters of a cycle of seven monads in the same plane. The radius of a heptagon, when the side is unity, is 1·152, and the reduced distance of equilibrium by the assumed laws is nearly 1·143, and with an added central monad, 1·107. The distance of equilibrium, with an ether monad in the centre, will be intermediate, or 1·12204 nearly. But the distance of equilibrium for an ether monad itself is $\sqrt[6]{2} = 1·12\dot{2}45\dot{8}$, or almost exactly the same with the former. It follows that an ether monad will be in steady equilibrium in the centre of such a cycle, but not one of matter; so that their affinity for material monads must be greatly lessened by the repulsion of axial ether monads, which they will retain in steady and permanent union.

The union of nitrogen with hydrogen seems to have exactly the characters which would result from these mechanical conditions. Carburetted hydrogen is resolved only by a very strong heat, but ammonia is completely decomposed by passing through porcelain tubes, heated to redness. On the other hand oxygen and hydrogen have no compound of a gaseous form, or without an entire change of their separate properties by intense and violent combustion. So axial particles, joined with a cycle of six, will be *in equilibrio* on each side of the centre, and have a strong affinity to the radial particles. With cycles of eight, they will be either unfixed and disconnected, or exert a strong disturbing and decomposing power. But with cycles of seven, they will neither admit, as in the former case, of very stable union, nor exercise, as in the second, a strong disturbing power. They will replace, with some difficulty, ether monads already occupying the axis with a relation of stability, and being united in a less stable manner themselves, any considerable heat, or atomic oscillation, may be expected to separate them again.

The electric spark, in a mixture of hydrogen and nitrogen, produces no union, while hydrogen and oxygen explode violently. In ammonium the union of the two elements is so feeble, that it can only be maintained in amalgam with mercury, under a galvanic current, and is resolved when the current is withdrawn.

Again, there is a difference between the compounds of carbon and nitrogen with hydrogen. One of carbon combines with two of hydrogen, and conversely, and there is a large number of hydrocarbons in which the number of the atoms differs by one only. But the chief compound of nitrogen and hydrogen is ammonia NH_3, while the two hypothetical compounds, amide NH_2 and ammonium NH_4, never exist in a separated form.

Now a double cycle of seven, with three monads in the axis, will have some degree of permanence, or be in a state of imperfect stability. One monad will be midway between the cycles, and one on each side. But the limits of stability being so narrow, from the near equality of the radius to the cohesive limit, a change in the compression or expansion of cycle, by union with other atoms, may enable two to rest *in equilibrio* in the planes of the cycles, or two in the planes, and one also on each side. These effects follow naturally from the slight excess of the radius above the neutral limit, and its exact coincidence with the enlarged limit for ether atoms.

79. Nitrogen, in contrast to oxygen, does not sustain combustion, and, on the other hand, can scarcely be called combustible. Its compounds with oxygen are peculiar. The Protoxide, by its density as gas, is oxygen 8 + nitrogen 14 = 22. Its Binoxide is lighter, and answers in density to oxygen 8 + nitrogen 7 = 15. The former is a general supporter of combustion, the latter only in special cases. Hyponitrous is $NO_3 = 38$, nitrous acid $NO_4 = 46$, and nitric acid $NO_5 = 54$. The last is a powerful oxidizer, and its presence plainly gives its explosive power to gunpowder, and to similar compounds of saltpetre.

Let us now conceive two single cycles, one of seven,

the other of eight monads, to be associated together. Their radii of equilibrium will plainly be different, the cycle of seven being less remote from the axis, as 1·12 to 1·26, or 8 to 9. But this difference will be lessened by their union, which tends to contract the one, and expand the other. Their numbers being prime to each other, there can be no tendency to a fixed angular relation, but they must act on each other nearly as two continuous circular wires, and their axial distance be nearly $\sqrt{1 - \cdot 14^2} = \cdot 99$, or very slightly less than the neutral distance. Any increase of *vis viva* will expand the larger cycle, while the other will be hindered from expanding by the central ether, and thus the oxygen will easily combine with those substances which have an affinity for that element.

Again, let us conceive one cycle of seven to be included between two of eight monads. The contraction will be greater than in the former case, from the greater number of non-adjacent monads. The second cycle of eight, by its attraction for the first, will bring them nearer than the distance in the last structure from the middle cycle, which will be repelled inward, and slightly contracted in size. Hence, in case of combination with a combustible body, there will be three causes of difference. A stronger connection will have to be overcome. The oxygen cycles, for the same quantity, will have less atomic *vis viva* to part with, and one portion of this will be absorbed by the re-expansion of the nitrogen, when the new compound leaves it free and separate. Accordingly, the binoxide supports combustion in some cases only. "Burning sulphur and a lighted candle are extinguished by it; but charcoal and phosphorus, when in vivid combustion, burn with increased brilliancy."

These explanations, of course, are in some degree provisional, and may have to be modified, when the theory is more fully developed, and applied by the help of a strict analysis, to determine the relative stability and inherent *vis viva* of different cyclical combinations. It would be premature to attempt their extension to conjectural arrangements for the more complex compounds of oxygen and nitrogen. It is enough to justify a general confidence in the present theory, that it gives results, even at first sight, in singular harmony with the distinctive characters of these four leading elements, and of some of their simplest compounds.

80. *Carbon and Nitrogen.*

These two substances form one principal compound, Cyanogen, which has many of the properties of chlorine and other simple bodies. Its formula is C_2N. It is a colourless gas possessing a strong pungent and peculiar odour. Under the pressure of $3\frac{1}{2}$ atmospheres it becomes a limpid fluid. It extinguishes burning bodies; but is inflammable, and burns with a beautiful purple flame. Its density, that of hydrogen gas being one, is 26, or nearly double that of atmospheric air. Paracyanogen is an isomeric substance, but solid, with a probable composition, C_8N_4. Cyanogen has a remarkable tendency to combine with elementary substances, and with metals rather than their oxides. Fulminic acid, the basis of fulminating silver, is Cy_2O_2. Hydrocyanic gas is formed from equal measures of hydrogen and cyanogen, and like the similar compounds of hydrogen with chlorine, bromine, and iodine, the bulk is not altered by the combination; so that its atomic weight as a gas, hydrogen being one, will be $13\frac{1}{2}$, or a fractional value. Its properties are highly poisonous.

It decomposes easily under the influence of light, but small quantities of acids hinder the decomposition. Cyanogen is obtained at a low red heat from the bicyanide of mercury. The affinity of Palladium for this element surpasses that of all other metals.

Now let us suppose that cyanogen consists of two carbon cycles of six, and two nitrogen cycles of seven, blended into two composite cycles of thirteen monads. Its basis, then, will be another prime, distinct from 3, 7, those of its components, and from 2, the prime basis of oxygen. Being dual, its symmetry will contribute to its permanence. But in all the more subtle relations which have to do with odour and electric harmony, when compared with 6, 12, 8, 16, the usual and frequent cyclical numbers, or 7, 14, those of nitrogen, it will have the same analogies as a musical discord. It will combine with metals as a simple, like oxygen, chlorine, iodine, and bromine. But in combinations with compound atoms of low cycles, it will tend to resolve itself, by a nearly equal division, into its component nitrogen and carbon cycles. Its atomic number 26, and that of Palladium 53·3, have very nearly the ratio of 2 to 1; and this may possibly have some connexion with the strength of its affinity for that metal.

81. Thus a review of the leading characteristics of the four simplest elements, hydrogen, carbon, oxygen, and nitrogen, and of their simpler and more fundamental compounds, reveals in every case numerous and striking agreements with the *primâ facie* results of the present theory. These are the elements where, above all the rest, we might expect to find a key to the real nature of the atomic structure. Their wide diffusion, their integer ratios to each other, the specific characters and universality of water,

their simplest compound, all invite us to some explanation which shall strip their properties of an arbitrary, empirical character, and reveal them as the distinctive and natural results of their derivation from some simple and universal form of matter. But other elements, and many other classes of phenomena have to be examined, before the competency of the hypothesis can be fully tested, and a clear warrant provided for submitting it to a more profound analytical investigation.

CHAPTER VIII.

GENERAL RELATIONS OF THE CHEMICAL ELEMENTS.

82. THE Electrolytic order of the elements differs widely from that of their atomic weights. They may be thus distinguished into successive groups, as follows:

1. Oxygen, sulphur, chlorine, bromine, iodine, fluorine, nitrogen, phosphorus, selenium.
2. Arsenic, chromium, molybdenum, tungsten, boron, carbon, antimony, tellurium, tantalum, titanium, silicon, osmium, hydrogen.
3. Gold, iridium, rhodium, platinum, palladium, mercury, silver.
4. Copper, uranium, bismuth, tin, lead, cerium, cobalt, nickel, iron, cadmium, zinc, manganese.
5. Zirconium, aluminium, yttrium, glucinium, magnesium, calcium, strontium, barium, lithium, sodium, potassium.

The contrast between this electrolytic order, and that of atomic weight, is very conspicuous. Hydrogen and gold, the lightest and heaviest atoms, meet here in the middle of the list. Chlorine 36 and potassium 39, fluorine 19 and calcium 20, which are very near in atomic weight, are found at or near the two extremes.

On the present theory the atomic numbers must depend on the relative number of monads which compose each atom, and the electric order on their radial or axial arrangement. For the positive pole of the electric current must be conceived to represent a superior velocity of atomic rotation, or greater *vis viva,* and the elements which tend to that pole or electrode must be those which have the greater moment of rotation, or a less axial and more radial structure. Hydrogen, of the simple monads, must be neutral in this respect, or claim a middle position. The regular metals must be supposed, from their cohesion, to be slightly or strongly polar; and the precious metals which are least oxydizable, to come near to hydrogen with regard to equality of moments, which is indicated also in mercury by its great fluidity. Atoms of nearly equal weight, if they come near the extremes, may be expected to have contrasted properties, but when they come near in the series, to have a close resemblance. Thus cobalt and nickel, which are near iron in both respects, are also magnetic and hard metals. But chlorine and potassium, fluorine and calcium, which have their atoms nearly equal in weight, but at opposite ends of the scale, have a strong and peculiar affinity for each other. It is clear that a hollow atom, of few cycles and large radius, and one of more cycles and small radius, may combine radially by the contrast of their characters, and an intense evolution of *vis viva* or ethereal motion must result from such a union.

83. The gaseous and vaporizable elements, except hydrogen, occupy the foremost places in the electric scale. The exception is explained at once, if hydrogen represent the simple monad, which must be indifferent with regard

to axes of rotation. The others must answer to hollow cycles, maintained by centrifugal force in stable equilibrium, and therefore not easily susceptible of a solid structure. If we assume 16 or 2·8, 2·16 or 4·8, 2·18 or 3·12, 4·20, 6·21, 19 or 8 + 9, and 14 or 2·7 to represent the arrangement of these high electro-negatives, their electric order will have a probable explanation, along with their atomic weight, and there may be thus some near approach to a just definition. Such elements, by their structure, must be capable of combining radially with those of a small circuit, and will set loose central ether, and an answering amount of *vis viva* or atomic heat, by such a union.

84. The next class, in which the electro-negative character prevails, includes carbon, boron, and silicon, and nine acid-forming elements. Carbon has been already examined, and the nature of its atom conjecturally defined. Let us consider boron and silicon separately, and the other nine elements.

The atom of boron, according to Berzelius, is 5·5724, or doubling, 11·1448; but according to Turner, Fresenius, and Fownes, 10·9. The mean of these is the integer 11, which may claim to be the true value. It is a prime number, and its simplest division is 6 + 5, where 6 is the atom of carbon, and five a factor of calcium = 20, and probably of several other earth-forming metals.

The properties of boron are very peculiar. It is a dark olive-coloured substance, a non-conductor of electricity, insoluble in water, alcohol, ether, and oils. It bears intense heat in close vessels, without fusing, or any change but a slight increase of density. Its spec. gr. = 2 nearly. If heated to 600° it takes fire, and boracic acid is generated. Boracic acid is a solid hydrate, spec. gr. = 1·479, and is rather bitter than acid, reddens litmus paper feebly, but renders turmeric paper brown like the alkalies. All the borates, in solution, are decomposed by the stronger acids. Anhydrous boracic acid is a hard, colourless, transparent glass. One of the strongest affinities of boron is with fluorine.

The characters of boron are thus intermediate between those of carbon and the bases of the earths. Like carbon, it is a dark, unmetallic solid, and forms an acid by its union with oxygen. On the other hand, its oxide is solid and not gaseous, and has some resemblance to silicon, as a glassy substance, imperfectly soluble. Now if we conceive it to be formed of two cycles, one of six and the other of five atoms, it will be half carbon, half earth, assuming a cycle of five to be the typical character of the earth-forming bases. Again, it has a special affinity with fluorine, as shewn by the formation of fluoboric acid gas. But the two primes 11 and 19 are not only both exceptional to the series of composite numbers, but differ by eight, so that the larger may readily include the smaller, with a distance slightly exceeding the neutral distance between them. Such a relation will be one of strong cohesive affinity. On the other hand, the resolved cycle of six and five will explain the characters of boracic, as intermediate to those of carbonic and silicic acid, and the double relation of boron itself to carbon and silicon.

85. Silicon or silicium, in its oxide silica, is one of the widest diffused and most universal of substances. We should thus infer, on the present hypothesis, that its structure must be one of great simplicity. It is the base of an earth, but much higher than the other earth metals in the electric scale, so that silica has the functions of an acid,

and forms silicates of the alkalies and earth oxides. The element seems also to be dimorphous. "It is a dark brown powder, destitute of lustre. When strongly heated in a covered crucible, its properties are greatly changed, and it becomes darker in colour, denser, and incombustible."

The equivalent of silicon has been variously reckoned, by Thompson at 7·5; by Berzelius (Miller) 23·71, by Turner 22·5, by Fresenius 14·79, and in Fownes and others at 21·3. Their differences arise mainly from a varying conception of the composition of silica, whether it is a protoxide, dentoxide, or teroxide. According to these three hypotheses, and the two different estimates of the ratio of the parts in silica, the values will be 22·5 or 21·3, 15 or 14·2, and 7·5 or 7·1. In this diversity, and considering the uncertainty of the argument from the number of atoms in similar compounds, which is contradicted by the isomorphism of potass and hydrated ammonia, the liberty may be claimed of selecting the most probable value on internal grounds, or from the requirements of the present theory.

Let us assume, then, the true value to be 15 = 3·5, and we shall have three cycles of five for a probable form. This may plainly exist in two forms at least, as the cycles are exactly superposed, or the radii of the middle cycle alternate with the others, so that the height is diminished nearly as $\sqrt{3} : 2$. The radius of a pentagon is ·8506, and of inscribed circle ·688, the side being one. If two cycles were radially superposed, the particles being doubly intermediate, the radial distance of those at neutral distance from the two nearest monads in each of the two adjoining cycles will be, apart from the general compression, ·688 + ·707 = 1·395. But the unreduced radius for an octagon, or cycle of eight, is 1·307. Hence two cycles of eight, superposed radially to three cycles of five, will have a position not far remote from one of neutrality, or, in other words, of maximum stability.

Again, fluorine has a strong affinity for silicon, as for calcium. But if a cycle of 19 is brought near to a triplet of cycles of five, any common revolution must tend to decompose the former into two portions, nearly equal, of 10 and 9, the former of which will be susceptible of being superposed symmetrically to the others, while the nine will also be superposed, but so as to produce a double want of symmetry.

The density of silica is $2\frac{2}{3}$, which would answer to 24, if its atoms are compressed in the same degree as those of water. Its atomic number, on the present supposition, would be 31. Also, if a cycle of 5 be the typical feature of earthy bases, then silicon, if composed of three only, would be the simplest of them, and the highest in the order of electric arrangement.

86. The nine metals which belong to the same class, as most electro-negative, are these.

Arsenic 37·5 or 75, chromium 28 or 26·7, molybdenum 46, tungsten 92, antimony 64·5 or 129, tellurium 64, tantalum 92 or 184, titanium 24·1 or 25, and osmium 99·6.

All of these agree in some common properties. With oxygen they form the arsenious and arsenic, antimonious and antimonic, chromic, molybdic, tungstic, telluric, tantalic, titanic, and osmic acids. To these may be added vanadium, forming vanadic acid. Arsenic, antimony, and tellurium also combine with hydrogen, a very unusual feature in metals. Arsenic is distinguished by the poisonous properties of its oxide, and the quality of subliming without being capable of fusion.

All these features point to the general conclusion, that these metals have no axial or central monads, that they have only three cycles, or four at most, and two radial sets of monads at least in each cycle, and have thus affinities with oxygen, chlorine, &c. by their outer shell, and with hydrogen by the inner, or the want of axial monads, while their flat structure, or large radius compared with their height, renders them more susceptible of volatilization than metals in general. Silicon, in electric order, has the lowest place in this class, and perhaps it may be inferred that no cycle of less than six enters their composition.

87. The third and middle class, in electric order, consists of gold and silver, the precious metals, of mercury, the only fluid metal, and of platinum, iridium, rhodium, and palladium, all found in the same ores, each pair having nearly the same equivalents, and also of great hardness and density.

The equivalent of gold, in Berzelius, is 198·9, in Turner 199·2, in Fresenius 196·5, and Fownes 197, or the estimate varies nearly from 196 to 200. That of mercury is 202·5 Berzelius, 202 Turner, 200 Fresenius, 100 Fownes, where the frequent ambiguity appears between the single or double value, and if the larger be taken, the atoms of gold and mercury are very nearly the same. In like manner Platinum is 97·2 Berzelius, 98·8 Turner, 98·7 Fresenius and Fownes; and Iridium 98·8 Turner, 98·6 Fresenius, and 99 Fownes. Here the difference in the estimates of the equivalent of the same metal is greater than their difference from each other.

Again, Rhodium is 120 Berzelius, 52·2 Turner, Fresenius, and Fownes; and Palladium 112·6 Berzelius, 53

Fresenius, and 53·3 Turner and Fownes, a third instance of very near approach in the equivalents, as they result from the latest inquiries. Silver is reckoned by all authorities at 108, with hardly a variation.

The characters of these metals agree with the hypothesis of a somewhat cubical structure of their atom, with factors not very unequal, and probably rounded off, at least in mercury, into a nearer approach to the sphere. It is conceivable that the variation of one or two units in their estimated equivalents may arise from a real latitude in their structure, or that some of them exist in allotropic forms, slightly different, and depending on the presence or absence of one or two material monads, and their replacement by monads of attached ether.

88. Let us begin with a perfect cube, with six for its side, which would contain 216 monads. The pressure of the whole, excepting the corner monads, and any centrifugal force, will maintain the relief of the edges, and expose them to possible abrasion. Let us suppose the two terminal planes on each side to lose their corner monads, and the two central ones to retain them, the compound atom will consist of $216 - 4 \cdot 4 = 200$ component monads. If the corner atoms are again removed from one corner only of the two central planes, the number will be 198, if from two corners, 196. In other words, there will be an atom composed of four planes of 32 and two of 34 monads. Such may possibly be the structure of the atom of gold.

Again, let us take the cube with five for its side, or 125 monads. Let us suppose the middle plane unchanged, the two next deprived of their corner atoms, and the terminal planes reduced to a square of nine, with two atoms added symmetrically on each of its four sides. We shall then have a compound atom of $25 + 2 \cdot 21 + 2 \cdot 17 = 101$ monads. This regular bevelling of a form, fundamentally a cube, must evidently tend to produce fluidity, by destroying all polarity, and making it easy to vary the axis of rotation, and for the atoms to approach one another equally on every side. But 101 is the earlier estimate of Turner and Berzelius for the atom of mercury, and that metal is distinguished from every other by its fluidity at all common temperatures. A compound atom, such as has been just described, appears to satisfy these conditions.

89. The atoms of Platinum and Iridium, in Turner, are both 98·8, in Fresenius 98·68 and 98·56, and in Fownes 98·7 and 99. There is thus no well ascertained difference between them. Platinum is the densest of known metals, ranking in this respect higher than gold. It is also much more tenacious, and is nearly infusible. If we conceive five planes or cycles to be formed, each of two intersecting squares of four, or a square of 16 monads, with another monad symmetrically added on each side, we have a compound atom of 100 monads. Or again, if four such cycles have a square of nine added on each side, we have a symmetrical atom of 98 monads.

The atom of silver is 108, which might be formed by three planes, each a six-sided square, or by six four-sided squares, with a monad on each side between every pair of cycles, so as to give a nearer approach to a cylindrical form. Assuming, however, the truth of the general hypothesis, it is only by degrees, and after full investigation of the results of each arrangement, that we can reasonably hope to attain, for each element, a certain knowledge of its exact atomic form.

90. The next class includes the more oxidable proper metals, copper (63·3, 31·6, 31·7), uranium (217, 59·4, 60), bismuth (141·9, 71, 213), tin (117·6, 57·9, 58·8, 58), lead (207, 103·6, 103·56, 103·7), cerium (92, 46, 47·26, 47?) cobalt (59, 29·5, 29·5, 29·5), nickel (59·2, 29·5, 29·5, 29·5), iron (54·3, 28, 28, 28) cadmium (111·5, 55·8, 55·8, 56), zinc (64·5, 32·3, 32·5, 32.6), and manganese (56·9, 27·7, 27·6, 27·6), with their equivalents from Berzelius, Turner, Fresenius, and Fownes. Most of these decompose water at a red heat, and are oxidized. Iron is strongly magnetic, and cobalt and nickel come nearest to it in that quality, while bismuth is first in order of diamagnetic substances.

The natural conclusion, from the relative electric order of these elements, and their affinity for oxygen, is that they are more prolate in figure than the previous class; though the diamagnetic properties of bismuth and others may require, in their case, some modification of this view. There is also the serious doubt, in most of the metals, whether a single or double number is the truer value; especially since the best estimates for cobalt, nickel, and copper, have for their apparent unit the half hydrogen atom.

The most important of these elements is iron, and the integer value 28 is assigned to its atom by a general consent of the best analytical chemists. This has the same relation to nitrogen, which that of sulphur bears to oxygen, or is exactly its double. The wide diffusion of the metal, and its strongly characteristic properties, would seem to require a simplicity and symmetry in the structure of its atom, beyond those of rare occurrence. The number is composite, and resolves itself only in one way, or into four sevens. Let us suppose four cycles, each composed of a hexagon and one central monad, and we have plainly a very simple and coherent structure. The seven atoms in each cycle are in the densest possible arrangement, and would be all at the neutral distance, except for the slight cohesive force exercised on each radial atom by the three which are most remote. The horizontal axes will be equal, and slightly less than 2, and the vertical axis either 3 or 2·6, according to their vertical or alternate superposition. The atoms will thus be polar, strongly tenacious, susceptible of easy revolution round the central axis, by which they will assume magnetic properties. The nearest affinity will be with carbon, of which the cycle is six, and with which iron is known to have a very powerful affinity.

91. The last class, in electric order, includes zirconium, aluminium, yttrium, glucinium, magnesium, calcium, strontium, barium, lithium, sodium, potassium, or the metallic bases of the earths and alkalies. These are all strongly electro-positive, but the bases of the alkalies still more than those of the earths. Zirconium, yttrium, glucinium, strontium, barium, lithium, are comparatively rare; but aluminium, magnesium, calcium, sodium, and potassium, are substances, in their oxides, very widely diffused.

Now it is plain that, if radial or axial arrangement is the basis of the electric relations of a compound atom, a cycle of six forms a kind of limit between two opposite kinds of structure, since the radius and the side of the hexagon are equal. Cycles of five and four constitute two successive stages of a centrical arrangement, the converse of that which belongs to the gaseous and volatile elements.

One of these elements, which is most widely diffused, and which is almost like a link between the earthy and alkaline bases, is calcium, for lime has strongly alkaline properties. Its equivalent, by general admission, is precisely 20 times that of hydrogen. This suggests immediately the two alternatives of four cycles of five, or pentagons, or of five squares. The former may perhaps be assumed as its probable structure. Its oxide, hence, may be conceived as having a cycle of eight radially superposed between its middle cycles.

Again, magnesium has its atom reckoned variously at 25·34 Berz., 12·7 Turner, 12·62 Fresenius, and 12 Fownes. Assuming here the greater correctness of the earlier estimate, and taking the double value of Berzelius, we have 25 for the atomic number, or five times five. On this view Silicon = 3·5, Calcium = 4·5, and Magnesium = 5·5, would form a progression of simply constituted elements.

92. Sodium and Potassium are the metallic bases of the two alkalies, and come last in the electrolytic order of the elements. They are among the lightest solids, being both of them lighter than water, and are oxidized by it with strong combustion. Both their electric place, and the alkaline character of their oxides, imply in them some atomic feature which contrasts strongly with oxygen, chlorine, and the electro-negative sustainers of combustion. Now the character of these, on the present theory, is excess of latent *vis viva*, or a structure at once compressed at the circumference, and hollow in the centre, so as to admit, when the cycle is broken up, or combines with other atoms, of a much more compact and condensed arrangement. The opposite character must be a condensed structure, with repulsive force from the centre, and attraction laterally at the surface.

Let us consider Sodium first, as the element of smaller atomic weight. Its number, in earlier estimates, is 23·3, and in the later simply 23. Let us first conceive cycles of four monads vertically superposed. They will be *in equilibrio*, when the side of each cubic space is slightly less than the neutral distance. Now let four monads be introduced centrically into those spaces. They will be at the distance $\frac{1}{2}\sqrt{3}$ from all the enclosing monads, and the excess of repulsion $3\frac{1}{4}$ times the balanced repulsion at the neutral distance. Hence the planes must expand laterally, till they reach nearly the value $\sqrt{\frac{3}{2}} = 1{\cdot}225$, being hindered from vertical expansion by the cohesive force towards each other of the central monads. We shall thus have a compound atom of 24 monads, reversing the conditions of the gaseous cycles, an excess of repulsion between the centre and the edges being compensated by the attraction of the superficial monads, thrust out from each other beyond the neutral distance. In the absence of other matter, ether particles in excess must clothe the sides of such an atom, as they must occupy the line of the axis in gaseous cycles; and these opposite relations must produce a strong affinity. The properties will probably be nearly the same, if the terminal central monads are attached outside the first and fifth cycles, instead of within them. It is conceivable that there may be three allotropic forms in which there are four central monads within, two central and two attached, or two central only, in which case 23·3 would be the resulting mean value of the atom.

Another curious relation comes to light on this hypothesis. An atom of sodium, on this view, would have five spaces, each exactly resembling an atom of water, or would be $2\frac{1}{2}$ atoms of water completed into unity and symmetry

9—2

by an atom and half of hydrogen axially combined. Such a union may be conceived to be the result of strong and intense pressure. Now two facts are remarkable, that the salts of sodium are all remarkable for their great solubility; and the ocean and deep seas, where water is exposed to great pressure, all are impregnated with common salt, or chloride of sodium, so as to form one constant and invariable constituent. There is here a simple and natural explanation offered of this striking phenomenon.

The atomic number of Potassium is 39. Let us conceive eight instead of five cycles of four superposed, and seven axial monads introduced. We have then a structure precisely similar to that of sodium, but still more axial in its character, so as naturally to exceed sodium in its electro-positive character.

93. A striking fact in chemistry, which has caused much perplexity, is the isomorphism of the salts of hydrated ammonia with those of potass and soda, and the ammoniacal amalgam, of which one component, the ammonium, or hydurate of ammonia, cannot subsist alone. On the supposition that all our present metals are really simple, this fact is a great anomaly; and the conjecture of some, that nitrogen is a metallic oxide, merely increases the difficulty, for it is NH_4 or nitrogen with four atoms of hydrogen which answers to one anhydrous atom of potassium or sodium. On the present view, the analogy comes clearly to light. The nitrogen, in ammonia, may be conceived as two cycles of seven, with which three atoms of hydrogen are axially combined. The hydrate will add to these two cycles of four, and one axial atom, so as to form a compound atom, with two cycles of seven, two of four, and four central or axial atoms of hydrogen. But the

oxide of sodium may probably consist of two quaternions, superposed radially between the second and fourth constituent cycles. The number of axial monads will thus be the same. The cycles will be four in one case, and five in the other; and the radius of the cycle of seven and the superposed cycles of four, in the sodium, will have nearly the same ratio as the length of the two axes. But the central particles, in one case, will be inseparably joined by a strong repulsion, due to the contractile tendency of the fourfold cycles; while the other will separate easily, under heat, because of the expansive tendency of the cycle of seven, which is at equilibrium only when its radius exceeds the neutral distance. It seems unlikely that so many correspondences, growing out of the present hypothesis, even in its undeveloped form, should not be indications of its substantial truth.

CHAPTER IX.

ON STATICAL ELECTRICITY.

94. The Science of electricity, of late years, has received an immense development, and linked itself with every part of Physics, but its theory remains perplexing and obscure. Three different hypotheses have been proposed, and none of them can be viewed as satisfactory: the one electric fluid of Franklin, with the recent additions of Mosotti; the two fluids of Symmer, with the developments of Coulomb, Poisson, and Murphy; and the induction theory, or lines of forces, of Faraday and other eminent electricians. It will be convenient to state, first of all, some of the leading facts, and then to remark on the proposed explanations.

(1) All bodies, by friction or other means, are capable of receiving some modification of their surface, which is called an electric charge.

(2) The electric charge is of two distinct kinds, positive or vitreous, and negative or resinous, and when one surface is excited positively, the other is excited negatively.

(3) Surfaces charged with like electricity, repel each

other, and charged with unlike electricity, attract each other, with a force varying nearly or exactly, as the square of the distance.

(4) Some substances are conductors, and allow a charge to distribute itself easily over their surface, and others are non-conductors, along which it spreads very slowly and with difficulty.

(5) A charged surface tends to induce an opposite state on a surface opposed to it, or vitreous, to excite resinous, and resinous vitreous electricity. This tendency is exerted, not only across air or a vacuum, but solid insulating or non-conducting bodies, called in consequence dielectrics.

(6) Strong charges of opposite kinds neutralize each other through the air, or a conductor, by an electric spark, after which both bodies are neutral, or lose the signs of electricity.

(7) An electric charge diffuses itself over a conducting body, so as to be wholly at the surface, but densest at points, edges, or where the curvature is most abrupt.

(8) Electricity is developed in all chemical action, and may be thrown into a current, passing continuously between two opposite poles or electrodes. In this form it has power to decompose chemical compounds, and electro-negative elements appear at the positive, and electro-positive elements at the negative poles.

(9) Electric currents, in motion, attract each other, when they are parallel, at right angles to the line joining them, and repel each other, when moving in the same direction in the lines of the centres.

(10) Electric currents in motion, disposed in spirals, have all the properties of natural magnets, and natural or

artificial magnets share the properties of spiral electric currents.

95. The theory of Symmer, Coulomb, and Poisson assumes the existence of two electric fluids, each self-repulsive, but attracting the other, diffused in equal quantities through all matter, so as to be in a neutral state. These fluids are conceived to repel and attract by the law of the inverse square, but to have no direct influence upon the matter with which they are associated, and to be detained at the surface of conductors, either by the pressure of the air, or some cause not yet known.

The remarks of De la Rive shew the growing distrust of this hypothesis among scientific electricians. "We shall not discuss," he says, "the comparative merits of these two theories (of two or one fluids). The latter, in such sort as Franklin formularized it, cannot be admitted, we shall see presently for what reasons. The former, although subject to strong objections, is in the present state of science a very convenient and tolerably exact manner of representing this agent, that we term electricity. We may for the present say, it is very probable that this electricity, instead of consisting of one or two special fluids *sui generis,* is nothing more than the result of a particular modification in the state of bodies. This probably depends on the mutual action exercised on each other by the ponderable particles of matter, and the subtle fluid that surrounds them on every side; a fluid generally known by the name of ether, and the undulations of which constitute light and heat." And again, "What is this fluid, which undergoes these modifications? Here, we confess, are questions not resolved, and we are far from presenting the theory of the two fluids as the limit to our knowledge.

Further on, we shall see that electrical phenomena very probably depend upon the combined action of matter, and the ethereal fluid which fills the universe; and by thus approaching to Faraday's molecular theory we shall be nearer the truth, than with the hypothesis of two imponderable fluids, existing of themselves, and in a manner independent of bodies."

The objections to the theory of two fluids are palpable and numerous. First, it is complex and unnatural to suppose two such substances, alike in everything but in the sign of their action, diffused through all the universe in exactly equal proportions. Secondly, electrical phenomena are thus isolated from those of light and heat, with which they have the closest intimacy. Thirdly, the relation of electricity to ponderable matter is thus wholly unexplained. Fourthly, the excitement of electricity, or separation of the two kinds, under the hypothesis, is inconceivable. For their attraction, by the law of force assigned them, must be infinite when they are neutralized. Fifthly, their retention *in vacuo* is also unexplained. Sixthly, the universal presence of electrical excitement in chemical union and separation, the most important feature of modern chemistry, is also left without any solution.

The agreement of the two-fluid theory with experiment, in the few cases where it can be tested, is far from being so complete as some have urged. It has rather the air of an imperfect approximation, than a full solution. Thus (*Enc. Metr.* c. II. p. 139) the ratio of electricities on two spheres which have touched and been separated, when the radii are 1 and 2, 1 and 4, and 1 and 8, is by observation 1·08, 1·30, 1·65, and by the theory of Poisson 1·1601, 1·3168, 1·4443. The actual difference is considerable in both extremes. But, what is still more important, the law of change is quite different. The ratio of the extremes is 2 : 3 by experiment, and only 4 : 5 by calculation; and this experimental value, when the radii are as 8 to 1, is greater than the theory assigns to the case where the disproportion is infinite.

96. The Theory of One Fluid, as recast by Mosotti, has met with favour from other writers. It assumes three laws of action, all as the inverse square, of matter on matter, matter on ether, and ether on ether. The first and third are supposed to be repulsive, the other attractive, but the last two to have a constant slightly higher in its value than that of the first. It is assumed, further, that the molecules of matter have a solid and spherical form, upon which the ether presses; that the quotient of the repulsion of ether on ether at the unit of distance is immensely greater than its elasticity for the unit of pressure; that the elasticity varies as the square of the density; and that the relation of the density of the ether to the molecular actions is linear, and is the sum of the densities they would separately induce at that part of the medium.

These assumptions, it appears to me, are quite inconsistent with each other. A density varying as the pressure implies a repulsive force inversely as the linear distance. This law obtains for gases where the elasticity is due to a centrifugal force, but cannot possibly belong to a medium, in which the law of repulsion is the inverse square. M. Mosotti virtually assumes that distant parts of the ether repel each other by a law wholly different from that which he ascribes to each separate portion. The constant of force, also, in the distant action, is assumed to be immensely greater than for the contiguous particles, the force of which must chiefly determine the actual arrangement. Again, the supposition that the elasticity is as the square of the density contradicts the fundamental law of force, which infers a connection between them of a very different kind. For all these reasons the memoir in question, however skilful as a piece of abstract analysis, is in my opinion quite fallacious. The mathematical skill of the author has led his readers to overlook the inconsistency between the various assumptions on which the truth of the conclusion must depend. The explanation, also, of gravitation, by a slight difference between the positive and negative constants in two laws of force precisely similar, has a very arbitrary appearance, most unlike the simplicity and grandeur of natural laws. The admission of solid atoms, at the surface of which the repulsion is infinite, is a further departure from all the proved analogies of modern science.

97. The view of Faraday is widely different, and one of which it is not easy to form a very distinct idea. It lays down the principle that "an absolute electric state cannot exist in a body, but that every electrized body finds near it, either in insulating or conducting bodies, an opposite electric state to its own, which makes induction a general phenomenon." The general nature of the view seems to be the substitution of equal lines of forces radiating in all directions from electrical surfaces or magnetic poles, for definite central forces, decreasing by a fixed law; and the decrease as the inverse square is viewed simply as the result of the laws of space, or the fewer lines intercepted, when a body, finite in dimensions, is removed to a greater distance.

So far as this hypothesis would resolve electrical attractions and repulsions into secondary results, and not primary laws of central force, it agrees with the view to be presently unfolded, and only seems rather wanting in a clear explanation of the primary laws on which they must depend. So far, also, as it refers electrical or magnetical changes to the interception of lines of equal force, it agrees with the necessary consequence of mutual action in a highly repulsive ether. But when it seeks to dispense with the conception of central force altogether, even in the case of gravitation, it appears to me to reverse the true direction in which we are to seek for a clear and satisfactory key to the great problem of electrical and chemical change. The view here proposed will have at least this presumption in its favour, that it is almost intermediate between Faraday's view of molecular induction, and lines of force, and the earlier and more widely received fluid theories. It is time now to proceed to its direct development.

98. *To explain the presence of Free Electricity at the surface of bodies.*

The mean distance, in space, of the ether monads, it has been shewn to be probable, from the velocity of light, is not much greater, if greater, than the First Constant, or the distance at which two units, or double atoms of matter and ether, would be at rest by their one repulsion and two affinities. The force of gravitation, in these cases, is relatively so small that it may be neglected. The neutral distance, however, as reduced by the ethereal pressure, must in either case be less than the Ether Constant. Any number of units, joined together under the influence of this pressure, or not separated by uncombined ether, may be conceived to form a chemical atom. The ether, which penetrates solids and fluids, will isolate these chemical atoms from each other, and transmit the external pressure, so as to condense them also, while it will be modified in turn by their repulsion and cohesive affinity. The cohesive action between the different chemical atoms will be exercised, in part, on their component units or atoms, but more strongly on the monads of ether which lie between them, and which must be fewer in the dense metals, and more numerous in lighter bodies.

Every chemical atom, with its ether coating, thus contains in its essential structure a threefold bipolarity. It is composed of strongly attracting units or material atoms, and strongly repulsive monads of ether. When in rotation, it has a centrifugal action at its equator, and rest and attraction on the line of its axis; and its two poles have themselves opposite characters, determined by the direction of the rotation. The first has a close analogy with the phenomena of static electricity, the second with electrodynamics, and the third with magnetism. At the same time they are all most intimately related to each other. Centrifugal action agrees with attached ether in its repulsive power, so that one can mechanically replace the other; while the nature of the poles depends entirely on the same fact of atomic rotation.

The interior strata of such a solid body, below the depth where cohesive force is sensible, are under forces symmetrical on both sides. With a few superficial strata, and chiefly with the outmost, the case is different. On one side are strata of matter, and on the other mainly of free ether. This must therefore be condensed upon the surface by the whole amount of the affinity exercised by these outer atoms. The effect must be that, in contrast with the inner strata, they will all be positively electrized, using the term conventionally for a surcharge of ether. The greater part of the natural charge of each atom will be repelled to the inner side, and replaced by a larger surplus on the outer side. A few of the nearest strata will experience, but in a lower degree, the like change, which will be limited by the rapid decrease of the affinity for more distant ether.

The amount of this surcharge of ether must depend, in ordinary cases, on the structure of the body, the mechanical state of its surface, the general temperature, and the conditions for equilibrium of ethereal reaction in every set of opposed bodies. For each kind of body, at the same temperature, there will be a mean value or rate of ether thus attached to the unit of surface. Bodies in this state, though their surfaces are positive with reference to their interior strata, are neutral to each other, and will give no sign of electric excitation. But if by any means this mean or natural charge of ether is increased or diminished, a new and distinct class of phenomena will arise. If any separate units of matter are attracted to the surface, they will increase still further its condensing power.

99. *To account for the excitement of electricity by friction.*

This fact, in which the science had its origin, needs first to be explained by a true theory. The double fluid hypothesis throws no light on it whatever. On the present view there is a simple explanation. When two unlike surfaces, in their usual state, are pressed on each other, the fact of the pressure shews that their surfaces are brought within the distance at which the ether charge of each surface can exercise a strong repulsion. But this implies also a sensible action of each surface on the ether charge of the other. These three new forces, when the two surfaces are chemically or mechanically different, must cause an altered distribution of the two charges of ether between them. One will gain, and the other lose. And this change must extend to the surfaces themselves. Those which receive an excess of ether will have their distance from the lower stratum increased, and probably assume also a more polar, in contrast to an equatoreal arrangement, while the other surface will undergo in both respects an opposite change. Such surfaces, if capable of retaining this new state, when removed laterally, will have a positive and a negative electricity.

100. "*Polish increases the tendency of a body to acquire vitreous, and elevation of temperature its tendency to acquire resinous electricity.*" (De la Rive, I. p. 13.)

These results flow at once from the theory. Increased temperature, since it implies a greater oscillation or rotation of the atoms, must diminish their retentive power over the ether, and disengage some portion of it. Again, a polished surface must be more nearly a true plane than one unpolished. The action will become that of the whole surface, instead of some prominences only, and the tendency must be, that the polished surface will attach a larger share of the double charge.

A similar result follows, when two bodies of the same kind are rubbed together, a large part of one over a smaller part of the other. The larger surface is plainly that which is drawn away from the other. The pressure, before the motion, will drive out part of the common charge to the edges of contact, where it will be retained by the double attraction. The moved surface, when withdrawn, will attract part of this excess, which will be replaced from the parts in contact, and through these, from both the touching surfaces. Thus the moved surface will obtain an excess of ether, and be positive, and the unmoved one be negative.

101. *To explain Electrical Conductivity.*

This property forms one main electrical contrast between different substances, and links the science directly with the chemical properties and structure of bodies.

So long as a charge resides on the surface of a body, it must imply a change in the outer strata of the body itself, and chiefly on the outmost. This may be of two kinds; its elevation by a kind of capillary action to more than its mean distance from the strata below, and a varied arrangement of the poles, in the case of rotation, so as to become more or less receptive of ether than before. In a body positively charged it seems clear that the poles will be more turned to the surface, and in one charged negatively, the equators, as a more repellent position. Now if we suppose that, in non-conductors this change of the poles is more decided, and not capable of lateral transfer without great difficulty, and that in conductors there is either expansion without change of poles, or an easy transfer of the polarity, the contrast in their electrical properties is at once explained. In one case the charge may be compared to a land-flood over a level surface, and in the other to a series of lakes, embosomed in hills.

The metals are all good conductors, though the degree varies. The diamond, mica, glass, sulphur, jet, amber, resin, and gum lac, are non-conductors. All the last, except the diamond and sulphur, have a small or moderate density, and chemical atoms of a rather high degree of complexity. Hence it is clear that the bulk of the space which belongs to each chemical atom, and which may be occupied by ether, is relatively great. Again, the diamond, it has been shewn to be probable, has a purely octahedral structure, which implies a cup-like form for all the spaces between the atoms which form its surface. The metals have precisely an opposite structure, and probably, in most cases, have terminal planes, forming a level kind of surface.

102. *To account for Electrical Attraction and Repulsion.*

The fundamental law is thus expressed. "Surfaces charged with like electricities repel, and with unlike attract each other, and the force of attraction or repulsion varies directly as the product of the two intensities, and inversely as the square of the distance."

The diminution of force as the inverse square is here no proof of a direct law of central force of that particular kind. For the intensity of light varies by the same rule, which results, in that case, from the properties of space, when motion is propagated from a given centre. The law in this instance is secondary and derivative, not primary and fundamental, and the same explanation may apply to the attraction and repulsion of electrized bodies. The same view seems to result also from direct experiments. For Sir W. S. Harris has shewn that "the attraction of electrized bodies depends wholly on the form of their opposed sides, and not on that of the rest of the body. Two cones opposed by their base attract just as much as two circular discs, equal to their bases, and two hemispheres as two spheres of the same diameter (De la Rive, I. p. 68), and the attraction between two circular discs, one larger than the other, is the same as between

10

two surfaces equal to the smaller." These results seem wholly adverse to the theory of two electric fluids with central forces of attraction and repulsion, emanating in all directions. In this case the charge on the averted sides would have a full share in the total action.

Let us now consider the natural results, on the present hypothesis, of the modification of the surfaces of bodies just described.

Since the repulsion of the ether diminishes by a high inverse power, as the twelfth, it must be insensible at very small distances. The increase or diminution of the ether on the surface can thus produce no sensible attraction or repulsion, depending immediately on the special law of repulsive force. But since the elasticity of the medium is very great, and its pressure amounts to eighteen billions of pounds per square inch, it is eminently suited to transmit impulses, or *vis viva*, from one surface to another, whenever it is disturbed from absolute rest.

The ether of the atmosphere, however, can never be in a state of simple rest. For besides the annual and diurnal motions of the earth, which must tend to disturb it, the air has evidently its particles at a distance from each other at least eight or ten times their own size, and constantly changing their position by a rapid motion, which causes their repulsion or elasticity. Hence disturbances must be constantly propagated through the ether in all directions, even if due to this cause alone.

Besides the statical pressure, which the free ether exerts on all the bodies it surrounds, there will thus be a dynamical pressure, arising from the motion constantly diffused through it. The force thus caused by any local compression will plainly diffuse itself by the same law as the waves of light or sound, or its action on a distant part will be inversely as the square of the distance in the same line, while the nature of the disturbance will enter into the law of lateral or angular diffusion.

Let us now suppose that the effect of an increased ether charge, with its change of the poles of the matter, is to increase the elasticity, or the completeness and directness of the rebound, when free ether impinges on the surface. The body will then exercise an increased repulsion on all sides. But since there was equilibrium before, it will still continue, the change being equal in all directions.

Now let a second body, freely suspended, like the first, undergo a similar change. Let the repulsive force of the first be $1+e$, and of the second $1+f$, compared with the mean value. Impulses which fall on the first are returned in the proportion of $1+e$, and those on the second in the proportion $1+f$. Hence the secondary impulses which have been reflected from both surfaces, will have the ratio $1+e+f+ef$ to their nominal value. Of these terms, the first is the normal repulsion itself; and the second and third balance the two extra repulsions on the outer sides of the two bodies. But ef is a term of the second order, which remains without compensation, and measures the relative repulsion of the two bodies. It is positive when e and f have the same sign, and negative when they are different. It follows that there will be a positive repulsion between two similarly charged surfaces, and a negative repulsion or attraction between unlike surfaces thus opposed. The force also will diminish as the inverse square of the distance, but only for the surfaces which directly face each other, and in other cases will diminish rapidly, not only with the distance, but with the mutual inclination.

103. The general law of force, resulting from this view, will agree with that derived from the fluid theory in depending on the two intensities and the inverse square of the distance, and in being repulsive for like, and attractive for unlike charges. But it differs in two important respects. The force will diminish rapidly with the inclination of the surfaces, as well as their distance, and it will be destroyed, or wholly modified by the interposition of solid matter. This corresponds perfectly with the general conclusions deduced by Sir W. S. Harris from his experiments.

Again, it has been seen that when spheres of radii 2, 4, 8, are separated after touching a sphere of radius 1, the changes are found to be 1·08, 1·30, 1·65, while by the fluid theory they are 1·16, 1·31, 1·444. Now it is plain that the present theory, which makes the force diminish, not only with the square of the distance, but the inclination, will tend to increase the relative influence of the larger sphere; and thus its values will differ from those deduced by the fluid theory in the same direction as the experimental values.

"Sir W. S. Harris has shewn that, when the charge is feeble, the increase of force is not exactly as the square of the quantity, but varies more rapidly. Thus in one experiment, when the quantities were doubled, the repulsions were as 1 to 5." (De la Rive, p. 67.) This result is easily explained on the present view. The increase of ethereal elasticity is not the same thing as the quantity of surcharged ether. If we assume that, in weak charges, a portion is latent, and does not act sensibly at the surface, the effect here noticed would ensue. If one fifth of the smaller charge were latent in each case, the active ratios would be $4^2 : 9^2$, or just one to five, as in the experiment.

104. The question still remains, in what way a surcharge of ether can act in varying the modulus of elasticity for the surface, or increasing or diminishing its repulsive energy towards the free ether which impinges upon it. It results from the laws of mechanics, and the assumed constitution of the atoms, that no *vis viva* can be destroyed, but can only change its form. But it seems plain that if the external ether impinges upon revolving equators, instead of fixed poles, it will partly modify their rotation, and partly be repelled tangentially, or along the surface. If this explanation be correct, the continuance of negative electricity on the surface of a body is connected with the production of an answering amount of atomic heat in the body negatively electrized. It agrees well with this conclusion, that collodion and gun-cotton are the extremes in the list of negative electrics. For these are both explosive substances, the latter in the highest degree, and their atoms must therefore be naturally receptive of insensible atomic heat in the same proportion.

105. *To explain Electric Induction.*

"When an electrized body is presented to an insulated conducting body, signs of electricity are developed, even though the bodies are a greater or less distance apart. These signs disappear as soon as the electrized body is withdrawn. This constitutes electricity by induction." (De la Rive, p. 82.)

This phenomenon results directly from the nature of electrical action here supposed. For the excess or defect of dynamical pressure on the side opposed to the electrized surface must alter the arrangement of the ether or electrical charge. It will dispose itself in such a way that the lateral pressure from the difference of the ether charge at different parts of the surface compensates the difference of external or dynamical pressure, arising from the pressure of the electrized body.

106. *General Law of Distribution on Conducting Surfaces.*

In the fluid theory, as developed by Coulomb, Poisson, and Murphy, &c. the self-repulsive force of either fluid is supposed to drive it to the surface, where it is detained by the pressure of the atmosphere. The law of its distribution depends then on the shape of the surfaces, being uniform in a sphere, but unequal in every other figure. Some simple cases have been calculated by the theory, with the help of a rather high analysis, and the results agree moderately well with the experiments of Coulomb. The general conclusion is that the accumulation is greater where the curvature increases, as at the end of the minor axis of an ellipse, and is infinitely great at points or angles.

On the other hand, Sir W. S. Harris concludes from his experiments (*Brit. Ass. Rep.* 1847, p. 24) as follows:

"First, that the influx of a charge is uniform, like the filling of a vessel with inelastic fluid, like water, and not with elastic fluid, like air. Secondly, that the distribution is equal, apart from induction, so that the intensity of a charged rectangle is the same as when rolled into a cylinder, and of a circular area, the same as in a sphere of the same surface. He infers that if a single charged body were to exist alone, there is no reason to suppose an unequal distribution. Thirdly, that the whole action of two opposed spherical surfaces depends on the distance of each pair of corresponding points, and may thus be reduced to a very simple formula, depending on their radii and mutual distance.

The contrast between these conclusions, and some of the results of other experimenters, seems to prove the need of further inquiry, before the facts can be viewed as thoroughly ascertained. Estimates of electrical force require peculiar accuracy and care for their determination. But the conclusions of Sir W. S. Harris agree in substance with the consequences of the present theory.

First, it seems incredible that each point of two electrized surfaces acts only on one point of the other, singling it out by a kind of elective affinity. On the other hand, a calculation based on this principle will approach to the case where the action diminishes with the inclination as well as the distance, more nearly than the complex calculation from fluids of equal force in all directions. It will not be difficult to solve the simplest cases, with an assumed law of decrement for the inclination, and to shew that they approach nearly to Sir W. S. Harris's empirical rule.

Again, the present theory agrees with the conclusions of Sir W. Harris, in viewing the influx of a charge as resembling more nearly the influx of water than of air. On the theory of two fluids, the whole internal space of the body is equally receptive of the fluid, and it is driven to the surface by its self-repulsion alone. On this view, also, the higher intensity at the minor axes of an ellipsoid is due to the repulsion of the whole charge, acting in straight lines through the substance of the body. The present theory, on the contrary, supposes that the change resides on the surface, because it is only at the surface that the relations of the matter and ether are discontinuous, and the distribution resembles the case of a soluble gas, admitted into a space of very small height over a large liquid surface. The self-repulsion of the charge will be insensible for all sensible distances; and hence its distribution, apart from secondary action or pressure, will be sensibly uniform, so far as it depends on central forces alone. But the equilibrium required between the lateral pressures and the retaining force will cause it to be denser where the curvature is greater, and most of all at edges and angles, though in a less ratio than in the fluid theory.

107. *Particular Cases of Electric Distribution.*

The statements of De la Rive and Sir W. Harris, compared, answer to the conclusions that appear to flow naturally from the present hypothesis, though a full comparison would require complex calculations, and a reference to the conducting or coercing power of different substances.

(1) "In the Ellipsoid the charge at the ends," according to De la Rive, "is proportional to the axes."

(2) "Those plates, whose length is at least double the breadth, have the charge nearly constant, till about an inch from the end. It is double at the end, and if the proof plane is put in the prolongation of the plate, fourfold." The retentive force on the ether, at the edge, has to counteract two equal pressures from the two sides in the same direction, and must therefore, it would seem, be double the force along the side of the plate. At the corners, for a like reason, it would be fourfold, and the proof plane held in the prolongation of the plate, is under the same condition, and will naturally receive a fourfold

charge, that is, an amount of ether causing a fourfold resistance or tension.

(3) "In a circular plate, it increases slightly to an inch from the edge, at one third of an inch is double, and at the edge is triple." The resolved part of the pressure of the edge which meets and balances the pressure on each side is $\int_0^{\frac{\pi}{2}} \sin\theta\,\delta\theta$, and the total pressure $\int_0^{\pi} \delta\theta$. Hence it would seem that the charge at the sharp edge will be π or $3\frac{1}{7}$ when the charge along the surface is unity; while its diminution inward will probably vary with the size of the plate, and the conducting or coercing power.

(4) "In a cylinder $33\frac{1}{2}$ inches long, and 2 in diameter, the charge being unity in the middle, it is $1\frac{1}{4}$ at 1·8 from the end, and $2\frac{3}{10}$ at the end." The charge at the circumference of the end, where there is a right angle, it seems a probable conclusion, from the theory, and the balance of pressures, should be $\frac{\pi}{2} : 1$, compared with that along the cylinder. But the end being small compared with the whole surface, this charge may be diffused over the whole end, and thus require the same ratio at the circumference to balance its own increased pressure. The extreme value would then be $\frac{\pi^2}{4} = 2{\cdot}46$, as its limiting value when the radius of the end is small, compared with the length. At a distance from the end equal to the radius, it would probably be $\frac{\pi}{2} = 1{\cdot}57$, and hence a value 1·25 at the distance of 1·8 seems to agree with this approximate result of the theory.

(5) "In twenty-four equal spheres in contact, it is

nearly constant for the middle ones, and 1·75 for the extremes." Such a series evidently approaches to the case of a long cylinder with hemispherical ends. In this case, we should expect a double strength of the charge at the ends, to meet the opposite pressures, reducing itself to unity a little beyond the first hemisphere. Hence the mean charge of the spheres next to the ends may be expected to be 1·5, or nearly a mean between the limiting and central values; and the mean charge of the last spheres a mean between this and the limit, or 1·75.

Thus a first and rough application of the present theory to the best ascertained facts of electrical distribution yields results equally, and perhaps more conformable to experiment, than the fluid theory. At the same time, since it recognizes the double influence of induction and the conductivity or coercive power of the substance electrized, and requires us to distinguish the quantity of ether distributed along the surface from the elasticity or electric force which it occasions, it plainly admits of two elements being introduced into the formulæ of electric distribution, by which the harmony between theory and experiment may be rendered more complete than in the fluid theory, which takes no account of the varieties of coercing power, and makes the charge a direct measure of the external activity.

CHAPTER X.

ON THE ELECTRIC CURRENT.

108. The Science of Dynamic Electricity, while it has made immense progress through the labours of Volta, Wollaston, Davy, Oersted, Ampère, Cumming, Becquerel, Faraday, De la Rive, and many others, still remains in a state not a little perplexing and obscure. Electricity, chemical affinity, magnetism, light and heat, are all proved to be intimately related to each other, and the relation of each pair of them supplies a large class of phenomena, but the exact nature of each and all continues unknown.

Electro-dynamics, or the theory of the electric current, and of the attendant chemical union and decomposition, stands first in order among these closely related branches of science. And here the unsolved questions are many. What is the meaning of an electric current? Is it the transfer of one or two fluids, properly electric, or of the luminous ether, or a transfer of forces alone? Why does it decompose chemical compounds, and some, not others? Why should such currents circulate, as the hypothesis of Ampère assumes, around the atoms of magnetic bodies? Why should electricity, in motion, have a power to attract or repel, which ceases when it is in repose? Why should one current cause another for a single moment, when formed or broken? Why should two solid and one liquid conductor be usually required? Why should heat alone produce such currents? Why should the same elements, joined in pairs, have constantly the same electric character, as relatively positive or negative? Why should gaseous elements occupy one end of the series, and earths and alkalies the other? These and many similar questions seem to baffle, while they stimulate, the curiosity of scientific men.

109. Two rival views struggled long for victory in the general conception of the voltaic circuit, the theory of contact, and the chemical theory. The latter has now prevailed, through the labours of Faraday and De la Rive; but the part fulfilled, either by the liquid, or by the union of the two metals in the circuit, remains very obscure. Dr Turner states the view of Davy as follows, and suggests a modification.

"Sir H. Davy considered chemical substances to be endowed with natural electric energies; meaning thereby that a certain electric condition, positive or negative, is natural to their combining molecules; that chemical union is the result of electrical attraction taking place between oppositely excited atoms; and decomposition from combined atoms being drawn asunder by electric energies of other atoms, more potent than those by which they were united. He regarded the poles of a voltaic circuit as two centres of electrical power, acting by repulsion on particles of the same electric state as itself, and by attraction on the opposite. Substances which appeared at the + pole were termed electro-negative, and those electro-positive which were separated at the − pole....If he meant that a particle of free oxygen or chlorine is in a negatively excited state, this is contrary to the fact. If sulphur unites with oxygen because it has a positive energy, why should it unite with potassium, which is far more positive than itself? The only way in which these facts seem reconcilable with the theory, is to suppose all bodies in their uncombined state electrically indifferent, but that they have an appetency to assume one state in preference to another. Electro-negative bodies are such as assume negative excitement under a certain approach to others, which at the same time become positive, chemical union being the consequence. On this view it is intelligible that sulphur may be positive to oxygen and negative to potassium. The following view seems best to harmonize the facts and the laws of electricity. A particle of zinc and one of oxygen, possessed of positive and negative electricity, assume, in combining, opposite electric conditions, and combine in consequence; adhering together by virtue of their opposite states, as two oppositely excited balls are mutually attractive. The zinc particle, on becoming positive, gives off negative electricity to the mass of zinc to which it belonged, and the particle of oxygen, on becoming negative, supplies positive electricity to adjacent particles. Thus electro-positives, in combining, give out negative, and electro-negatives positive electricity." (Turn. *Chem.* ed. 7, p. 128.)

Again, De la Rive states the views of Ampère and Berzelius, and his own latest conclusions.

"When we study these phenomena, we are forced to admit a simple relation between the atom and electricity. Ampère had supposed that each atom of matter possesses an electricity proper to itself, positive or negative, and that in the state of equilibrium it is surrounded by the contrary electricity, which disguises the former. This hypothesis, which explains elegantly a certain number of facts, is open to grave objections. It does not explain how the same atom can be sometimes positive, at other times negative, according to the atom with which it is in relation. Berzelius admitted that each atom has two electric poles, one positive and one negative. But to this simple hypothesis another was added, not at all probable, that atoms are unipolar, or keep one of their electricities in combining, and abandon the other....I am disposed to admit in the atom a natural polarity. All the facts relating to the development of electricity, particularly by heat, seem to lead to this conclusion. With respect to the objection, that the atom being naturally spherical, there is no obvious reason why it should have polarity in one direction rather than another, we have only to suppose that each atom originally received a motion of rotation upon itself, and we obtain an axis and direction of rotation, and a different pole at each extremity." (Vol. II. p. 48.)

This view is unfolded more fully in a later passage.

"The principle from which we set out is, that every atom has two electric poles, contrary and of the same force. Whether this is due to a movement of rotation on itself, or to another cause, is of little importance; it is with us a primitive fact. One atom differs from another, only inasmuch as it has a more powerful polarity than the other, but in the same atom the two poles are of the same force. When two atoms are brought near, they attract by their opposite poles; but as they are spherical, they can come in contact only by one of the poles of the former, and the contrary pole of the latter. We shall be obliged to admit a principle which, it seems to us, is founded upon a great number of facts; that when two atoms are free and insulated, the positive pole of that which has the stronger polarity unites with the negative pole of that which has the more feeble polarity. If they have the same force of polarity, there is no reason why they should unite by two contrary poles, rather than the other two. Then they are not attracted by their poles, but simply obey the molecular attraction; and this is the case of homogeneous atoms, or of cohesion; whilst, when the atoms are heterogeneous, they are attracted by their opposite poles, and then obey chemical affinity. This is therefore the result of the attraction of two differing atoms by their contrary poles; so that the positive pole of the more powerfully polar is united with the negative pole of the other. The compound atom has equally two poles; and these are equal, because the excess of the more powerful over the more feeble of those united neutralizes part of the electricity of the free pole of the more powerful: the compound atom is therefore found under the same conditions as the free atom. Thus chlorine will unite by its positive pole with hydrogen, and by its negative with oxygen. But if we place the compound atom between contrary polarities, it will be so arranged that its + pole is turned to the negative, and its – pole on the positive side, of the same pole....The difference between the natural state of a liquid, and its electric polarization is this, that in the former state the molecules are in complete electric equilibrium, and the equal and contrary electricities with which the two poles are endowed, unite by the surface; while in the latter state the equilibrium is broken, the contrary electricities no longer uniting by the surface, but the positive poles are all turned on the negative side of the pole, and the negative poles on the positive side." (Vol. II. pp. 875—7.)

110. None of the views of these five eminent men is probably without a partial truth. But their divergence shews the difficulty of the subject, and the cloud which still rests upon it. They all share the common objection, that they make electricity consist of two distinct, non-convertible fluids, one positive, and the other negative, strangely combined, and as unaccountably separated, when so many facts imply that the two states are simply correlative, since the same substance may be positive to one and negative to another substance. But each of them is liable to further difficulties.

First, the view of Davy lies open to Dr Turner's objection, that free, uncombined elements shew no trace of either electricity, and are sensibly neutral. And next, the analogy fails in the main point. For when opposite electrics attract each other, the attraction ceases with the discharge; while in chemical union there is no apparent discharge, but abiding attraction. Again, the two fluid theory, in dealing with static electricity, assumes their distribution to be independent of their action on matter. But the view of Davy requires us to assume the reverse, that their attraction for matter is more powerful than for each other, and that this superior attraction finds, in every element, a distinct and graduated amount of one fluid only.

The remedy Dr Turner suggests for one of these faults introduces another still more serious. For what reason can be given why two neutral elements should each part with one of the neutralized fluids on the mere presence of the other? Why should a separation, reversing the assumed laws of electric attraction, take place in both at the same moment? How can we account for the one attractive force of the two elements, by assuming a separation in each of them, with no force at all, of two electricities mutually attractive, and previously united? The difficulty thus introduced seems twice as great as that which it attempts to remove.

The view of Ampère agrees with that of Sir H. Davy in ascribing to each element a proper or constituent electricity; and both are in this respect, perhaps, nearer the truth than Dr Turner's modified explanation. But when, on the principles of the two-fluid theory, he supposes the opposite electricity to be present in equal amount, and to disguise the first, he falls back into the original difficulty. The true conception of a proper electricity is that it is self-disguised or latent by the structure of the atom. But if it requires to be neutralized by an equal amount of the other kind, we return to the point from which we start. We have merely the standard case, in the fluid theory, of matter with two neutral electricities, with an unexplained propensity, in each pair of elements, to part with opposite fluids in presence of each other; so as to create, by this double sacrifice, effected without any force, a mutual attraction, which is assumed to last, against all analogy, when their combination has made them neutral once more.

111. The later and riper hypothesis of Prof. De la Rive introduces a new and important element, when it suggests that the atoms are essentially polar, and probably through a movement of rotation. This idea is a great step towards a full solution of the problem. But the suggestion is half retracted in the second passage, where the polarity is called an ultimate fact, of which the precise cause is not important; and the further assumption, that the atoms are

spherical, robs the first of its chief value, while the explanation is defective and imperfect in other ways.

First, the polarity of rotation is of a definite kind. It implies two poles, equal and opposite, on the two opposite faces; but these equal polarities are plainly inseparable, and can never neutralize each other. Two atoms, polar in this sense, could have no excess of positive or negative polarity, but must be positive on one side, and equally negative on the other, and the one which is the more positive must also be the more negative. Hence a gradation of positive and negative electric character, if positive and negative are defined by right-hand and left-hand rotation, would plainly be impossible.

Next, the different degrees of polarity, in the hypothesis of revolving spherical atoms, could consist only in different rates of revolution. Now there may be plain mechanical reasons why two revolving atoms, either of equal or unequal velocity, should unite with unlike poles together, so that the direction of motion may concur. But there is no conceivable cause why one pair of unlike poles should combine, rather than the other, the relations of each pair, both of common direction, and equal or unequal velocity, being just the same. Positive and negative, when applied to the opposite sides of such atoms, are convertible terms; and there is no such difference as could explain the supposed law, that the north pole, for instance, of the faster should combine with the south pole of the slower atom, and the south pole of the faster never combine with the north pole of the slower.

Thirdly, the supposition that the excess of negative electricity of the stronger atom would be neutralized by the surplus of the inner or united pole, seems to be an illusion from the ambiguity of terms. If neutrality consists in the equality of contrasted motions on the opposite sides of the atom, then the rotation must be equalized before it can be true of the compound atom. But if the opposite electricities are conceived to be separate and in polar contrast, in the simple atoms, then the difference of rotation in the two inner poles will have no effect to neutralize the excess of the negative pole in the compound atom. Two different senses of the term positive, as applied to like poles of rotation, or to a greater or less amount of rotation, seem to be confounded together.

Fourthly, the assumed law of union, besides its strangeness in itself, since it seems to imply a choice without a determining reason, appears inconsistent with the presumed course of electrolytic action. Calling the north pole positive, or that on the side of which the motion is the reverse of the hands of a watch, it is assumed that the north pole of the oxygen joins the south pole of the hydrogen in each atom; or if the line be north and south, and the revolution the same way as that of the earth, then each atom of oxygen will be south of that of hydrogen, and the free north pole belongs to the hydrogen and the free south pole to the oxygen. When the terminal particles are withdrawn, and the nearest unite, it is plain that the order will be reversed, and the south pole of the oxygen in the first atom combine with the north pole of the hydrogen in the second. No subsequent reversal of position, for the electrolysis to be renewed, can alter the fact of this change. In fact, the view of electrolytic action, adopted from Grotthus and Roget, clearly implies the perfect indifference of the pairs of differing atoms, on which side they are united together. For these reasons, besides others to which Professor De la Rive himself alludes, the explanation he has offered, though perhaps a nearer approach to the truth than those of his predecessors, appears to labour still under a serious and insuperable defect.

Let us now return, to consider what light may be thrown on this very difficult subject by the present view of the atomic structure.

112. *To explain the diverse Polarities of the Chemical Atoms.*

Every chemical element, on the present view, is not a solid sphere, either in motion or at rest; but consists of a definite number of units, duads, centres of force, or monads of matter and ether inseparably joined together, arranged in some definite order, revolving usually round some axis of rotation, and parted from the nearest elements by a certain amount of attached ether. Their free ether, again, is supposed to consist of a finite number of monads, not very disproportioned either in their distances or number to those of the material units, or resembling less a fluid atmosphere than a large number of satellites. Again, the pressure of the free ether will be transmitted throughout all bodies by the perfect elasticity of the monads and units, and will condense and compress every chemical atom, so as to isolate them from each other. Their cohesive affinity, it will result from the facts of science, and the assumed laws, will be considerably less than this ethereal pressure; but it will operate equally in spite of this disproportion, just as the atmospheric pressure, being equal within and without, leaves the texture of all animal tissues unimpaired.

Every such element, then, must include in its very

structure four or five different sources of contrast, from which a kind of polarity may arise. First, there is the contrast between the duads or units of matter, which attract only beyond the first or neutral distance, and the monads of attached ether, which are repulsive to ether at all distances. Secondly, there is the contrast between a larger or smaller charge of ether to the same element, giving rise to a polarity of quantity and not of reverse position, and hence admitting of graduation. Thirdly, there is the possible repulsion of the ether to one end or side of the atom, so as to clothe it unequally, another contrast which admits degrees of intensity. Fourthly, there is the contrast of a swifter or slower rotation, admitting also of various degrees. Fifthly, there is the contrast of radial or axial structure, admitting of a few distinct gradations alone, from the most centrical to the most open or cyclical form. Sixthly, there is the contrast between north and south poles of rotation, or oppositely placed hemispheres, and this is one of complete equality, but incapable of degrees, or of neutralization. Seventhly, there is the contrast between the plane of the equator and the line of the poles, or two rectangular axes, along which there is a special repulsion, and a third, along which there will be a special attraction of double the amount. All these contrasts may be included under the general term of polarity. They are also linked together, in the hypothesis, by mechanical laws. But they clearly differ very much from one another, and their combination will account for the natural perplexity which has arisen from the constant appearance of polarity in all the phenomena, and the complex and almost inconsistent features this polarity assumes.

113. The first kind of polarity or contrast, which lies at the basis of the science, is the distinction of vitreous and resinous, or positive and negative electricity. In the theory of Franklin, this depends on the excess or defect of one single electric fluid; but in that of Symmer, Du Fay, Coulomb and Poisson, which has been more prevalent, on the pressure of two opposite fluids, exactly equal in quantity and similar in properties, but self-repulsive, and mutually attractive, and differing in sign only. This view makes it impossible to give any explanation how one fluid can be separated from the other, or any difference in their affinity for any atom exist between them.

On the other hand, there are many facts which agree with the hypothesis that excess and defect in the quantity of the same fluid or cause of force is the true contrast of positive and negative excitation. Thus heat increases the tendency of a body to become negative, and increase of motion would plainly throw off part of the attached ether by common mechanical laws. "The greater expansive force of positive, compared with negative, electricity for equal tensions is established by a great number of phenomena" (De la Rive, II. p. 883). Again, the neutral state, on this view, would not be absolute and invariable, from the very nature of two fluids, but will depend on the special class of relations considered, whether the exchange of ethereal pressure between solid surfaces, the friction or contact of different surfaces pressed together, or still closer atomic approach in chemical combination. Accordingly, we find that elements which are neutral in reference to distant and general action, are strongly positive and negative in their atomic approach, as oxygen and potassium. The fact, also, that their electric relations, as positive or negative, throw the chemical elements into a series, with gaseous elements at one end, and the bases of the fixed alkalies and earths at the other, proves that the contrast is not like that of two faces of rotation, or two opposite parts of an indefinite straight line; but resembles rather the difference of higher and lower temperature, and approaches to a real zero, or absence of ethereal activity or expansive force at one end of the scale. If we assume positive electricity to consist in an excess, and negative in a defect of the mean amount of attached ether, the main features of contrast will be the same as on the other view; but the possible change of this mean under different relations will account for many phenomena otherwise inexplicable, and elements at the extremes of the scale will evidently have, as experience confirms, a wholly different constitution.

114. The phenomena of the Electric Discharge agree with this view of the two electricities. Its velocity in copper wire, by Wheatstone's experiments, is 288,000 miles per second, or one half greater than that of light in space. But the tension of the ether, in solid substances, must be increased for direct and rapid impact, and the velocity, like that of sound in water, be greater than for transverse undulations in the planetary spaces. Yet the relation is so near to equality, as almost to prove that there is a direct ethereal impact; as when a series of equal ivory balls, in a right line, are struck by a first ball, the last will detach itself with a nearly equal force at the other end.

115. A second kind of polarity or contrast is that which appears in the directive power of an electric current. When it passes horizontally from north to south, a needle

below it has its north pole turned to the east and a needle above it to the west. But when the positive end of the current is to the south, and the negative to the north, these directions are reversed, and so in any other position. We have here a strict polarity, without graduation, which is satisfied at once, by supposing the directive power to depend on the direction of the movement, and the atomic rotation. The plus and minus, in this case, are independent of the force of the current or the limit of neutral electricity. They depend simply on the two opposite directions of the current, supposed to be of a single fluid or ether, and in one direction alone, not in both directions at once, as in the double fluid theory.

116. A third contrast is that of native electricity, or the electro-motive order of the elements. There is here a regular scale, from oxygen at one end to potassium at the other, though liable to deviations and partial inversions according to the nature of the fluid in the circuit. Still, there is a plain distinction in the character and properties of the beginning, middle, and end of the scale.

Here we have the signs of a polarity or contrast, intermediate to the two others. It is not a simple antithesis, for the same body may be positive to one element, and negative to another. It admits therefore of a certain amount of graduation. On the other hand, it does not seem capable of indefinite extension, like a positive and negative quantity, but has a fixed limit, apparently, at each end, beyond which no element more positive or more negative can be found.

These features, which seem inexplicable in the case of spherical atoms, find a simple explanation on the present view. They correspond to the varieties of axial and radial structure, in compound atoms composed of a definite number of units and monads. These are not a simple contrast, like the two poles of rotation, but admit of several degrees. At the same time they have a limit, each way, in the most cyclical or most axial structure consistent with permanence, and the actual conditions of molecular structure. Thus oxygen, nitrogen, fluorine, at one end of the scale, are gaseous elements which cannot be condensed; chlorine condenses with difficulty, and sulphur, iodine, bromine, phosphorus, and selenium, are volatilized with ease; while the earths and alkalies are basic, and almost unsusceptible of being turned into vapour. An axial structure, which implies a low rate of rotation, or comparative rest, is most favourable to the attachment of ether, and a rapid motion, causing repulsion, is most unfavourable; so that the normal state of the bases may be called positive, and that of the opposite elements, negative, by their very structure; while they are externally neutral, or in a mean state, until their relations are changed by contact and union with each other. Thus the fundamental idea of Davy and Ampère will be retained, that some substances are naturally negative, and others positive, but will be cleared from an addition, borrowed from the fluid theory, which obscures and destroys its true meaning. This natural electricity does not need to be neutralized by a second fluid, but is the direct result of the atomic structure, and leaves the substance neutral, whenever its atomic structure is not altered by union or separation. Again, the objection which Dr Turner has brought against Davy's view is removed. For a swiftly revolving atom must be *in equilibrio*, with regard to external pressure in general, like one at comparative rest, or of axial form; and still its absorptive or negative character will be apparent, as soon as its rotation is diminished by union or even contact with other atoms.

117. A fourth contrast, not abrupt but gradual, is that between conductors and non-conductors or dielectrics, on which so many phenomena depend. Liquid conductors seem to have a middle character, and to be capable of conducting by electrolysis alone. On the present view, it depends on the ease or difficulty with which attached ether can transfer itself from one atom to another, surmounting the resistance of that transmitted ethereal pressure, by which they are parted from each other. Now it is plain that, in metals, from their higher density, a larger space will be occupied by the nuclei, and less by the attached ether, and a stronger cohesive force traverses the interval, so that on both accounts the transfer of ether is more easy. Again, it seems plain that the ease will be greater where there is an approach to a cubic or spherical form than where the length or breadth is unequal. Accordingly, silver, copper, gold, the best conductors, are found very near the middle in the electric series of elements.

Again, the polarization of dielectrics will answer to the condition of the atoms, when there is a strong retentive or coercing power in the attached ether, but it is disposed unequally on opposite sides of the element, from the charged state of neighbouring atoms. The contrast, again, between a swifter and slower rotation, in connection with the varied structure of the atoms, will link itself with the properties of conductibility for heat, thermo-electricity, and the development of heat in chemical combination.

118. *To define the nature of the Voltaic Current.*

The main conditions on which the Voltaic Current

depends are the contact of two different metals, and their union by a fluid conductor, acting chemically upon the more oxydable metal. When there are many cells, and the extreme plates are joined by metal wires, the current is closed and active, or broken, as the wires touch, or meet in a liquid conductor, or else are withdrawn from each other. There is thus, included by the air and non-conducting cells, a circuit within which positive electricity travels one way and negative electricity the other; and the strength of affinity, by which the zinc is oxydated in the cells, measures the strength of the electrolytic power between the terminal wires, when placed in water or some other fluid. The intensity of the electricity, also, increases with the number of cells; but the heating effect is greater with single plates of large surface, and the deflecting power on the needle depends on the quantity, and not the intensity of the current. Its velocity varies with the nature of the wire and lateral induction; but the results of experiment place it, when highest, at three fifths of the velocity of light, and it seems to be retarded by induction to one fiftieth only of that velocity. The contest between the contact and chemical theories seems now decided in favour of the latter; but still the fact requires to be explained why the presence of two metals, as well as a process of oxydation or chemical change in one of them, is almost essential to the transmission of a current, or at least the accumulation of electric power.

Now, first, since by the hypothesis the ethereal pressure permeates all bodies, and serves to isolate the chemical atoms from each other, a limited series of conductors, where the ether can pass with comparative ease from one atom to the next, need to be in equilibrium with regard to all these atomic pressures, and the resulting distribution of the attached ether. When no force is at work in any part of the circuit, which exceeds the coercing power of atoms on their own outmost ether monads, there will be no electro-motion or current. But when this force acts at any part, so as to cause a new distribution, there will be a new condition of equilibrium propagated throughout the circuit; and it is plain that the disturbance by chemical union in one part of the circuit will be most simply neutralized by an answering amount of disunion or decomposition on the opposite side.

Let us now conceive two diverse atoms in this circuit to be brought near together; one with more rapid rotation, but a smaller amount of attached ether, the other revolving slowly or nearly at rest, but charged positively, in the same proportion, with repulsive ether. They will have a strong tendency to combine by their mutual affinity. Three or four effects must ensue. The swifter revolution will be lowered, and the slower increased, that the compound atom may assume a common velocity. But the pressure which separated them having ceased by their union, a corresponding amount of repulsive force must be set free, and dispose itself, under the form of ether, released from the positive atom, to find a new equilibrium through the circuit. But since the rotation of the negative atom has been destroyed to that amount by the union, the immediate defect of equilibrium is on that side, and the released ether must set out in that direction. On the other hand, the increased rotation of the zinc in combining, will tend to diffuse itself on the other side, and produce a loosening, to the same amount, of the attached ether. Thus a repulsive force, equivalent to the separating pressure, which is removed by the union of the two atoms, will travel round the circuit, on one side, in the shape of transmitted ether, and on the other in the shape of increased atomic motion. When the reunion is effected by metal wires alone, without an electrolyte fluid, the metal will resume its normal state, intermediate between the extremes of accumulated ether and increased rotation, and the surplus pressure or repulsive force will manifest itself in the form of sensible heat, or the repulsive action of the atoms, in contrast to that of their component monads. But when an electrolyte is interposed, the positive pole will attract the negative, and the negative pole the positive element, nearest to it, while the tension or pressure will thrust away the other halves of the two compounds, which will either reunite in the path of the circuit, or be propelled onward till they reach the opposite poles. The view of Grotthus, which supposes a decomposition and recomposition of all the atoms in the route of the circuit, seems open to the objection, that these changes could not begin till the electrolytic force has spent itself in dividing the terminal atoms, and there is no power left by which the separated oxygen or hydrogen atom could repel the like atom already combined from its actual union. The chief transporting power will be from the positive towards the negative pole, in the direction in which the ether travels; and this change has been established by a series of curious and delicate experiments. (De la Rive, II. pp. 424—442.)

The effect, then, of the completing of a voltaic circuit through good conductors, and in one limited direction, may be compared with that on a vessel of water, supplied by a siphon from a reservoir nearly level with its brim, and in which there are many minute apertures, through which the water oozes slowly, when a hole, rather less than the siphon, is opened near its base. The supply increases with the more rapid efflux, and the downward current thus formed diminishes the literal escape by the small orifices, so that the current is almost wholly in the new direction. Before the circuit is closed, the surplus ether of the chemical union must find some outlet, either by imperfect and slow conduction, or by conversion into atomic heat; and the reaction thus caused will hinder the rapidity of the action itself. But as soon as the circuit is formed, an outlet is provided for the ethereal pressure, which enables the chemical affinities to exert their full power in bringing about the new combinations.

119. The two main conditions on which the formation of the voltaic current depends are the chemical action of the liquid on the oxydable metal, and the contact of two different metals. In the explanation of Volta, the chief influence was attached to the latter; while in the chemical theory of Fabroni, Wollaston, Faraday, and De la Rive, the main cause is held to be the chemical action. But perhaps in this view, which has replaced the other, the office fulfilled by the contact of the two metals has hardly sufficient prominence, being, in all ordinary cases, no less essential than the chemical agency itself. On the other hand, the opinion that mere contact can be an unfailing source of electric power is disproved by facts, and does violence to one of the plainest mechanical laws.

On the present view the explanation of the effect of this contact seems very simple. It is a determinant of the direction in which two opposite forms of repulsion shall travel, excess of ether, and increased rotatory motion. By the hypothesis, the more negative metal, or copper in the usual pair, has the greater *vis viva* of rotation. Hence copper and zinc in contact, or soldered together, must both suffer a constraint, and depart from their normal state at the points of contact. The copper will have less, and the zinc more atomic motion. There will thus be a tendency of the copper to receive more ether, in compensation of diminished rotation, and of the zinc, to part with some of its ether, because its rotation is increased. At the same time, this increase of rotation predisposes it to unite more readily with oxygen, of which the rate of motion is assumed to be still higher. Thus the chemical change alone supplies the electrolytic energy; but the atomic contact of the oxygen with the zinc, and of the zinc with the copper, have an equal part in deciding the distribution of the liberated energy under two distinct and opposite forms.

CHAPTER XI.

ON ELECTRO-MAGNETISM.

120. An electric current, passing between the poles of a voltaic battery, has the directive power of a magnetic needle. When the current passes above the needle from north to south, the north pole of the needle is turned to the east; and the direction is reversed, when either the direction of the current, or the position of the needle is reversed also. A magnet, in its turn, acts on a moveable electric current, and tends to make it assume a transverse position. Electric currents, again, when disposed in spirals, act on each other and on magnets like natural magnets, or like poles repel and unlike attract each other. The action of the earth, also, or its directive power on natural magnets and on such artificial magnets, formed of spiral currents, is exactly the same.

These phenomena, first opened to view by the discovery of Oersted, and more fully developed by Ampère, have given rise to two or three slightly different explications. The most widely received is that of Ampère. He assumes that magnetism consists in a special arrangement of the atoms of magnetic bodies, by which electric currents constantly circulate around them. From the date supplied by experiment with regard to the mutual action of currents, three laws are deduced; that two currents moving the same way, parallel to each other, and at right angles to the joining line, attract each other; that, moving oppositely, they repel with equal force; and that, moving in the direction of the joining line, they repel with half the force. From these laws is deduced the action of closed or open currents on each other; and nearly all the varieties of magnetic action are also explained, if it be admitted that circular or spiral currents are moving constantly, in a certain specified direction, around their molecules or atoms.

Others have conceived of the relation between the current and magnetism in a different way. Oersted himself seems to have supposed that the conflict of two electricities, in the current where they meet, gives them a spiral direction, so as to act on the magnet at right angles to the direction of the wire. Mr Barlow, also, has reduced the whole theory to one single assumption, that "every particle of galvanic fluid in the wire acts on every particle of the magnetic fluid in the magnet, inversely as the square of the distance, but with a tangential force, which tends to place each at right angles to the other, and to the line of junction."

121. So far as this statement assumes two different fluids, one galvanic, the other magnetic, it is plainly more cumbrous than that of Ampère, which admits an electric fluid alone. But viewed merely as an expression of the mechanical relation between the magnet and the current, it seems to present it in its simplest form. On the other hand, it does not seem to explain directly the mode of action of one electric current on another.

The view of Ampère may claim, in two respects, a

superior simplicity. It dispenses altogether with a magnetic, as distinct from an electric fluid; and refers consistently to one common principle, the action of a current on a current, a current on a magnet, a magnet on a current, of the earth on magnets or currents, and of two natural or artificial magnets on each other. But its gain in these respects is partly balanced by a serious loss. The three laws from which it starts are not simpler in themselves than the one law assumed in the other view, and have just as little the appearance of being primitive laws, and not secondary results of some still undetected cause. The assumption that one or two fluids, by the mere fact of moving in certain directions, gave rise to three laws of angular force, by which they act on the same fluid, but only when in motion, is harsh and violent. Again, as Mr Barlow has observed, "it is an immense demand on the reader to request him to admit an infinitude of infinitely small currents, perpetually circulating within the solid mass of magnet or a steel bar." We may go further, and say that the assumption contradicts the definition of an electric current, as deduced from the only cases of which we have knowledge. For the current, either in the earth, or in the voltaic battery, plainly requires distinct masses, of a countless number of atoms, capable, by distinct chemical qualities, of receiving an opposite excitation. To transfer it to the ultimate atoms themselves reverses thus one essential condition, on which its formation in all known cases depends; and if the atoms are further assumed to be solid spheres, as Ampère and most others have done, the contrast amounts to a direct contradiction. To justify the hypothesis, we should have to assume chemical union and separation constantly in progress on opposite sides of the magnetic atoms themselves, or else currents circulating constantly with no possible cause of excitation.

A true explanation cannot wholly supersede either of these diverse, but nearly related theories, which have succeeded equally in grouping together large classes of facts, and in assigning the chief mathematical laws on which they depend. The test of its success must lie in reconciling them together, in freeing each of them from hypothetical elements, which disfigure them, and referring them alike to clear and definite mechanical laws, dependent on the very constitution of matter and ether. Let us see whether this important end will not be attained by the adoption of the present theory.

122. Since neither the atoms of bodies, nor the ether of space and of the air is in a state of rest, and they are linked together by a law of high repulsive power, it is plain that a continual transfer of *vis viva* must be carried from one part to another; and these changes must counteract each other in a state of equilibrium or permanent motion, like the transfer of caloric in bodies of the same temperature, on the old theory of a calorific fluid. If any kind of permanent motion is set up in this ether in a given direction, it must have some action upon all the surrounding ether. And if the motions of any part of this ether are linked with a certain portion of matter, itself moveable, the attractive, repulsive, and directive power will plainly be communicated from one to the other.

Whatever action of this kind occurs, since it consists in a transfer of *vis viva* through a perfectly elastic medium, it is plain that it will be inversely as the square of the distance, for elements having the same kind of motion, and the same angular relations, and differing in distance alone. At the same time, the dependence of this effect on angular position, and on the direction of the motion, will shew its derivative character, and distinguish it from a proper central force, like that of gravitation.

Again, since the force must result from small displacements, and these infinitesimal compared with all sensible distances, the effect must be linear, or the law of composition of forces must apply, and the total effect be capable of resolution into three directions at right angles to each other.

Again, the law of repulsion in the ether, the infinitesimal distance of its monads, and the great increase of repulsion on the least departure from the mean place of equilibrium, imply the existence of lines of equal force or repulsion, radiating from every monad to every other, however distant, and that the transmitted repulsion is equal, at whatever distance, along these lines.

The motion of any such monad is, in its effect on any other monads, as a source of *vis viva*, and not a mere centre of force, the transfer of one of these lines of force from a first direction to another infinitely near. It is measured, then, by the union of an attraction, represented by the first line, and a repulsion in the direction of the second, equal to the first when the motion is at right angles, but greater or less, when it inclines to or from the second monad, by the law of the inverse square. Hence, in the simplest case, the action on a monad at right angles to the direction of the motion will tend to move it in precisely the contrary direction, with a force inversely as the square of the distance.

If, however, the ether acting and acted on be grouped around material atoms, the motion will probably, however swift, be rhythmical, and depend on the intervals of the atoms, and in travelling the space of one such atom, act as a couple, being withdrawn on one side of the distant atom, and replaced on the other. It will thus have a direct force to turn the atom round an axis at right angles to the joining line, and to the direction of the first motion, so that the equatorial movement, on the near side, may be in the same direction with the inducing motion.

In general, the mechanical effect of any atomic movement of translation, transmitted through the ethereal medium to a distance, must be twofold, each inversely as the square of the distance. The part which is at right angles to the joining line is a force of initial rotation, backward with reference to the resolved part of the primitive movement. The other part is a repulsion in the direction of the joining line, when the motions meet, or are opposite, and an equal attraction, by the removal of a mean repulsion in the state of rest, when the motions concur or lie the same way. But for these effects to be manifest, or reveal itself in material motion, there must be two series or systems of material particles, one exciting, and one excited, which are linked by some fixed law of connection with the ethereal motions.

123. From these axioms, which seem to be necessary mechanical consequences of the assumed laws of repulsion and affinity in ether and matter, these results seem at once to follow. First, two parallel electric currents, moving the same way, will attract each other, since each will tend to revolve towards the other, in the direction of the slant lines, and the forward and backward tendencies in two equal, parallel lengths, will neutralize each other, leaving an attractive force alone. Secondly, two currents, parallel, and moving oppositely, will repel each other for the same reason. Thirdly, motion along the line of junc-

tion will also be attractive in its tendency, when both move the same way, and repulsive when the motions meet, so far as it depends on the general law of transmitted motion. But both of these theoretic effects will probably be insensible, because action in the very line of motion must depend on the primary law which decides the formation of the current, and on the conducting power, and not on a secondary effect alone, which could only be transmitted through the moving ether itself, and would cease, if the exciting motion be as swift as the ethereal transmission.

124. From the same principles, the relation between the electric current and the magnet, flows at once, without any need to invent electric currents, properly so called, circulating around all the magnetic atoms. To see this clearly, we have only to reflect on the nature of á magnet, and of the current, freed from hypothesis, and reduced to its simplest definition.

A magnet, then, has two poles, a north and a south, facing opposite ways, and like poles repel, and unlike poles attract each other. A revolving atom, in like manner, has two poles, a north and a south, if we name that pole towards which the motion reverses that of a watch, the north pole, and that on the side where the motion is like that of a watch, the south pole. By the general principle of the congruence and disagreement of motions, like poles of such atoms must repel, and unlike poles attract each other. The attraction and repulsion, also, depending on the greater or less amount of *vis viva* transmitted through the ether, when the motions of rotation concur or oppose, must vary as the inverse square of the distance. Every revolving atom, constructed as the present theory requires, is thus an elementary magnet by its very structure and rotation, without requiring the foreign conception of currents from without circulating around it. Any substance in which the atoms not only revolve, but are capable of being disposed with a large excess of poles directed lengthwise and on the same side, will be magnetizable, and if there be a coercive power to retain the poles, when once thus arranged, in their new position, it will have all the characters of a permanent magnet.

Again, while the relation between the two opposite faces of a revolving atom is plainly the same as between the poles of a magnet, the relation between the axis and the equator of rotation is precisely the same which exists between a magnet and an electric current. One is at right angles to the other. One has two opposite faces, according to the direction of the rotation, the other a direct movement of translation in the line of the tangent. By the principles of mechanics, an ether current, moving along in a medium with lines of equal force, will tend to produce a movement of rotation round an axis at right angles to its own course, and the joining line. In other words, it will tend to magnetize bodies that are magnetizable, by inducing rotations of the particles with the axes in that position. Conversely, where the body is magnetic, or a fixed relation established between the north poles, and the direction of the magnetic body, there must be an equal force exerted to turn the body into such a position, that the movement of these equators and of the current may agree together. Now this is precisely the law which defines the directive power that magnets and electric currents exert upon each other. We have simply to assume that the north pole in a magnet is that away from us when the motion is that of a watch, or which answers to the north pole of the earth itself, and then all the consequences agree with the supposition, the positive pole of the battery being assumed also to be the source of the ether current. Thus, when the current is from north to south, a magnet beneath it will have its equatorial motion southward, or the north will answer to the earth's west, the south to the earth's east, and by consequence, the east to the earth's north; or the north pole of the magnet must turn eastward, to produce mechanical agreement between the two directions of motion.

125. The phenomena of electro-magnetic induction seem to admit of a simple mechanical explanation by the same theory. For the first effect of a current newly begun, by the law of mechanics just stated, must be to produce a reaction in a parallel wire, depending on their distance and the energy of the first current. But this current, having no sustaining cause, must cease by the reaction of its own momentary effort, and the general inertia of the atoms, and the ether of the wire be in a new state of equilibrium. The cessation of the first current, being an equivalent change to the creation of an opposite current, will plainly have the same effect, and induce for the moment a current in the same direction as that which ceases. These are precisely the facts which have been brought to light in the last thirty years, and received so wide a variety of practical applications.

126. All the main features, then, of Electro-magnetism, the attraction of two parallel currents moving the same way, and their repulsion, when moving opposite ways, the main basis of Ampère's theory; the mutual influence of a current on a magnet, and a magnet on a

current, and of spiral currents on each other; the directive law, by which a magnet and a current place themselves at right angles, the one basis of Mr Barlow's theory; and the inductive power of a current, when first formed, and when broken again;—all find a direct and simple explanation, on mechanical principles, in the present theory. The fundamental law on which they rest is simply this, that permanent ethereal motion, in a medium where there is uniform repulsive force, or lines of force in all directions of constant value, tends to diffuse itself in all directions by the *vis viva*, to an amount inversely as the square of the distance; that like motions attract and unlike motions repel; and further that, by the couples which result directly from the motion, there is a directive force of similar amount, which tends to make rectilinear and equatorial movements concur in direction with each other. There is thus no need whatever to introduce supposititious electric currents, distinct from the magnetic atoms, and moving around them, to account for the phenomena. The rotation of the atoms themselves, which alone makes them polar and magnetic, is enough to explain the directive force exercised by magnets and by currents. The substance of the theory is retained, but imaginary vortices like those of Descartes, which only encumber it, are swept away.

127. This explanation requires, perhaps, one or two corrections to be made in the alleged laws on which Ampère has founded his theory of magnetism. The experimental cases on which the theory is founded are these:

(1) "Two equal and contrary finite currents exercise on a third, situated at the same distance from the two former, no action, the attraction and repulsion being equal."

This equal and contrary action results equally from the present conception of the mechanical effect, and is indeed almost self-evident in any view.

(2) "The action of a rectilinear current on a moveable conductor, is equal to that of a conductor bent and turned in any way, but between the same limits, the currents being of the same intensity."

This law requires, for its truth, that the sinuosities are small compared with the distance. In this case, neglecting quantities of the second order, it results evidently from the mode of mechanical action, and the equality of action in contrary directions.

(3) "A closed circuit cannot set in motion any portion of a current forming an arc of a circle of which the centre is on a fixed axis, around which it may freely turn, an axis perpendicular to the plane of the circle to which the axis belongs."

This seems to result at once from the nature of the mechanical force, in the present theory. For the motion of the exciting current acts on each element of the excited body only through the ether which it contains, and the reaction arising from the law of their union. Hence the only possible action in the direction of the element will be to increase or diminish the intensity of the current, or to have a slight accelerating or retarding power. The fourth case of equilibrium, between proportional conductors, similarly placed, resolves itself into the general law of the inverse square, and the general properties of space.

128. From this last case of equilibrium, however, the conclusion has been drawn by Ampère, and is given in Murphy (*Electr.* p. 113), that two portions of a current, moving in the direction of the line which joins them,

repel each other with just one half the force of their attraction, when they move parallel at right angles to the line of junction. This conclusion is drawn by an ingenious process of analysis, and it is inferred that this relation is necessary, in order that the total action on each element of the second current may vanish irrespective of the shape or distance of the first.

This reasoning, I conceive, involves a fundamental sophism, disguised by the ingenuity of the analytical process, and the apparent simplicity of the result to which it leads. Let us suppose any forces to act on a circular metal pipe, not directly, but by means of their action on a current of water which fills the pipe, and circulates within it. It is plain that they can only accelerate or retard the current, so far as they act in its direction, and can only tend to move the pipe, so far as they act on the current at right angles to its course, so as to produce a reaction on the upper or under, the nearer or further side of the pipe itself. But if the direct action on any part of the conductor in the direction of the current is null by the conditions of the problem, and can only result from the action in other parts of it, and the cohesion of the parts in a solid conductor, then the analysis is misdirected, and the supposed consequence fails.

Again, the law which results from this analysis has another presumption adverse to its truth, from the singular conclusion to which it leads. It implies that the attraction of one element on another is

$$\frac{1}{r^2}\left(\sin\theta\sin\theta'\cos\phi-\frac{1}{2}\cos\theta\cos\theta'\right),$$

where r is the distance, ϕ the angle of their two planes, and θ, θ' the angles of each element with the joining line. Assuming $\phi = 0$ and $\theta = \theta$, we have

$$\frac{1}{r^2}\left(\sin^2\theta - \frac{1}{2}\cos^2\theta\right)$$

for the attraction of two parallel elements which make an angle θ with the line of their centres. This implies a neutral state of complete equilibrium, when the elements make an angle of $35^\circ.16'$ with the joining line, an attraction at greater angles, and a repulsion at less. But it is hard to believe that two currents, whatever their precise nature, should be neutral and indifferent at this angle, and at no other, which is the necessary result of Ampère's formula.

Still further, the conclusion, which also follows, that two parts of currents, when moving in the line of junction, repel when the motions are similar, and attract when they are opposite, is in its own nature highly improbable. Every mechanical conception would point to an opposite relation, of attraction when the motions are similar, and of repulsion when they are opposite. The indirect evidence, from analysis, it has just been shewn, gives no real warrant for this paradoxical conclusion.

129. Two experiments are mentioned in De la Rive, I. p. 230, as confirming this law of repulsion of similar currents in the direct line. The first is when a part of the conductor, like a horse-shoe, rests on a bridge between two semicircles, dipping into mercury on each side, and on communication being made, with the fixed part of the conducting circuit, through the mercury, the horse-shoe recedes along the bridge. This is supposed to prove that each half is repelled by the current in its own line. But it is a simpler solution to suppose that each side is repelled

by the alternate part of the current, where the directions of motion are opposite, and not the same; or that there is direct ethereal repulsion on one side, and reaction on the other, at the points where the current enters and leaves the moveable part of the circuit.

Again, the motions caused in a capsule of mercury, through which a current passes by two ends of platinum wires, are supposed to prove the same law. But these, also, seem capable of a simpler and more natural explanation by the direct repulsion of the ether on the material atoms of the liquid conductor through which it is passing, which must be of the same kind as its heating effect in passing through solid wires.

All the main phenomena, then, of electro-magnetism appear to be solved, by the present theory, on mechanical principles, or the transmission of *vis viva* through a highly repulsive medium, and the consequent tendency to symmetry of permanent motions, whether of translation or rotation. There is no need to introduce a cumbrous hypothesis, like the vortices of Descartes, of countless electric currents, distinct from the magnetic atoms, and constantly circulating around them, without any assignable cause for their existence, and in the entire absence of one main condition, the contrast of two or three kinds of compound structure, on which that existence, in every known case, appears to depend.

CHAPTER XII.

ON MAGNETISM AND DIAMAGNETISM.

130. THE recent discovery of the division of all substances into magnetic and diamagnetic, or those which arrange themselves axially or equatorially between the poles of a powerful magnet, has opened a new and difficult field of inquiry. This is rendered more perplexing by the further discovery, still more recent, that magnetic or diamagnetic substances may be either in a normal or abnormal state; and that in the abnormal state, magnetic bodies behave as if diamagnetic, and diamagnetic as magnetic. The order of magnetic power is as follows, beginning with iron, the most powerfully magnetic, and ending with bismuth, which, in a lower degree, takes the lead in diamagnetic power.

Iron, nickel, cobalt, manganese, chromium, cerium, titanium, palladium, crown glass, platinum. Arsenic, ether, alcohol, gold, copper, silver, lead, water, mercury, sodium, flint glass, cadmium, tin, zinc, heavy glass, antimony, phosphorus, bismuth.

Among diamagnetics are also included rock crystal, many non-metallic salts, iodide of phosphorus, sulphur, resin, cooked or raw meat, blood, feathers, a piece of apple or pear. Among the gases, oxygen is strongly magnetic. These lists shew how widely the magnetic varies from the electric or voltaic order of the simple elements. Thus oxygen is magnetic, and sulphur and phosphorus diamagnetic; platinum and palladium magnetic, and gold and silver diamagnetic; iron, nickel, and cobalt magnetic, and zinc diamagnetic. The first of these are taken from the electro-negative, the second from the intermediate, and the third from the electro-positive elements.

131. The following explanation, by Professor De la Rive, is perhaps one of the latest proposed.

Magnetic bodies are those which have most atoms in the same space, and a low conducting power for electricity; and the diamagnetic have fewer atoms, or else a greater conducting power. In the isolated atoms the electricity, excited at one pole, passes by the surface to join the opposite kind, and returns by the axis. When a number of such atoms are disposed near together, in a circle, with like poles the same way, they form a molecular circuit, and have an electric current on their outer side. For this purpose it must be composed of atoms very near together. But "in order that the current may be formed round the molecule, it is not only necessary that the atoms be very near together, but that they be not sufficiently conductible to enable the two electricities, accumulated at their poles, easily to unite by their surface, rather than to unite with the contrary electricities of the two atoms, between which each of them is interposed." Thus copper and zinc, though with nearly as many atoms as iron in the same space, are supposed to be diamagnetic, because of their higher conducting power for electricity.

Several weighty objections seem to lie against this attempted solution. It is complex enough in itself, to assume, with Ampère, that electric currents are revolving constantly around all the chemical atoms; since an exciting cause, like that which results in the voltaic circuit from the chemical union of vast numbers of them, can have no existence in the case of the single atoms themselves. It is a further complication, to suppose that these spherical atoms, which determine the chemical constitution, are themselves composed of an infinity of smaller atoms, permeated by an electric fluid, so that a current can travel along the surface in one direction, and along the diameter in the other. It is a third degree of complication that these currents, by Ampère's hypothesis, besides continually separating and reuniting, without and within the atom, with no apparent cause, act on each other doubly, first as centres of force by the laws of static electricity, and secondly, by the mere fact of their motion, by rather complex laws. It is a further remove from simplicity, to suppose that these chemical atoms, even with all this complicated apparatus of external and internal currents, do not build up directly the structure of bodies, but must first be combined into molecules of an uncertain number of atoms, on which molecules alone the magnetic qualities really depend. A theory loaded with all these assumptions, however ingenious it may be, or whatever classes of facts it may partially explain, can have very small claim to be the true interpretation of the facts, or a faithful reflection of the simplicity of natural laws.

132. There are two other fatal objections which seem to lie against Professor De la Rive's solution.

First, if iron (II, p. 68) has 230 atoms in the same

space in which gold and silver have 150, the linear excess of atomic density will be as 6 to 7. But since this refers to the whole space which belongs to the atom of iron, and of gold or silver, or to the mean distance of the centres, the natural inference will be that the surfaces of the gold and silver atoms, and not of those of iron, are nearest together. For even if we double the usual atomic number of iron, still the silver atom has nearly double, and that of gold nearly four times its weight, so that the radii of their spheres, if equally dense, would have a larger ratio than 7 to 6. On the other hand, to assume uncertain varieties of density in these spherical atoms complicates the whole hypothesis still further. The presumption, then, must be, on this view, that the surfaces of the gold and silver atoms are nearer together than those of iron, which just reverses the required condition.

Again, the more conductive metals are supposed to be diamagnetic, in spite of their density; because this conductivity "enables the electricities, accumulated at their two poles, easily to unite by means of their surface, even when insulated, rather than to unite with the contrary electricities of the two nearest atoms." In other words, the conductivity assumed in the hypothesis is one which conducts electricity along the surface of the same atom, in preference to its passage from one atom to another. But the conductive power, which alone is revealed by experiment, and by which the metals are classed, is exactly the reverse, and expresses the facility with which the electricity passes from one atom to another, and thus can reach distant parts of the whole mass. A graduated conductivity along the surface of the same atom is a pure hypothesis, of which experience has told and can tell us nothing.

Hence both the causes offered for the diamagnetism of gold and silver, in the view now examined, would seem to imply the exact reverse, and prove them more magnetic than iron; because the surfaces of their atoms, if spherical, must probably be nearer together, and their greater conductivity from one atom to another would make the formation of the supposed molecular circuits more easy. These appear to me two fatal objections to the theory, besides the immense complexity of the atomic structure which it implies, and the total absence of any key to the nature and formation of these anomalous atomic currents.

134. Let us now endeavour to trace the results of the present theory.

First, every revolving atom, by its very structure, will be an elementary magnet. Since it resembles a cycle or a cylinder, more nearly than a sphere, it will have a natural axis of rotation, and its units, being a reunion of matter and ether, will have an action, as ether monads, on the attached and free ether; while the repulsion of the free ether, being everywhere the same, and in all directions, implies lines of equal force, which are variously affected by every atomic motion. Again, unlike poles, when they face each other, imply a concurrent motion, which they must tend to impress on the intervening ether, producing a radial repulsion at right angles to the joining line, and an axial attraction in the line, from a partial exhaustion of the ether, a cylindrical eddy, and imperfect ether vacuum. A magnetized body must be one in which the atoms rotate habitually, and where poles of the same kind are disposed, by a decided predominance, in the direction of the length towards one end.

Let us now suppose a body not magnetized, not even

naturally magnetic, to be placed between the poles of a powerful magnet. The law of assimilation of motion in the last chapter will imply, as its necessary result, that there will be a strong tendency to create a rotatory motion in the atoms of this body, agreeing with the rotation of the atoms in the direction of the two poles. If the atoms rotate already, the tendency will be to turn them, so that the chief amount of rotatory motion may lie in the line of the magnetic influence. If previously almost at rest, or simply oscillating, the tendency will be to make them rotate, so that the polarity may be the least disturbed; that is, around the main axis, if the ends have a strong polar attraction, and the sides are nearly indifferent; or at right angles, if the ends have only a slight, and the sides an angular polarity. In general, the greatest amount of concurrent motion will be when the length of the body is placed in the direction of the magnetic poles; and accordingly, the majority of bodies have, in this sense, magnetic properties, or a tendency to place themselves lengthwise, and not transversely, between the poles of a real magnet.

135. Three conditions, then, seem required in the chemical atom, to ensure a great magnetic susceptibility; an axial structure, so that the cohesion shall be mainly by the ends rather than the sides; the absence of lateral polarity, or a form nearly cylindrical, and the union of cohesive fixity of axis with a large moment of rotation. These conditions are, in part, opposed to each other, so that their union in the required proportions may easily be conceived to be rare. We may thus explain, perhaps, the singular fact, that only three of all the fifty or sixty elements have strong capabilities of magnetic action.

Any suggestion in detail must be only conjectural in

the present stage of the theory. Subject to this needful reserve, there seems an easy way of reconciling these conditions with the known properties of iron and its atomic number. This is given at exactly 28 in all the approved lists, and suggests at once an arrangement in four sevens as the simplest explanation. Let us conceive, then, four cycles, with one central, and six outer atoms, to constitute the element of iron. Apart from compression of the ether and the remoter units, and supposing the cycles vertically superposed, this atom will be a minute cylinder, of which the height is three times, and the breadth twice the neutral distance. Its sides will be nearly circular, or strictly hexagonal, and its ends will have a strong polarity. The ethereal pressure will force all the units of each cycle within the neutral distance, unless compensated by centrifugal force. Hence it seems easy to conceive a tendency to assume such a rotation, from the constant struggle between the pressure of the outward ether, and the developed repulsion of the component monads, as may keep the particles at the neutral distance, when the centrifugal force must just balance the ethereal pressure. In this case all the required conditions for a high degree of magnetic susceptibility seem to be fulfilled.

136. The magnetic power of iron is increased by heating it, up to a red heat. Beyond that limit it rapidly diminishes. This agrees perfectly with the present theory. For the first effect of increased temperature must be to increase the velocity of the rotation, and thus to add to the magnetic power, which depends on the axial rotation alone. But when the luminous limit is reached, the ends must be conceived to oscillate laterally, and the cohesion to diminish up to the point of fusion. Hence the magnetic

force must be weakened, and disappear, as soon as the atom begins to revolve on an equatorial axis, where fusion also begins. For though inductive magnetism may be shared by liquids and gases, magnetism proper is confined to solids alone.

137. Diamagnetism, in this view, depends on such a structure of the atoms as disposes their equators, rather than their poles of revolution, to arrange themselves in the direction of the length of the mass. The form which seems most favourable to this arrangement is the flat cylinder, of which the diameter exceeds the height. The same result may occur, in a lower degree, when the moments of rotation at right angles are nearly equal, and there is still a slight oblateness, or truncation of the ends, so that the equators predominate in the length, and the poles in the breadth of the substance. A third case may include those axial atoms, in which the length exceeds the breadth, but there is lateral polarity, and a less than usual adhesion at the ends. The first description may belong to diamagnetics, like arsenic and antimony and phosphorus, which are strongly electro-negative; the second to those intermediate in their electric order, as gold, silver, mercury; and the last to the electro-positive, such as zinc and sodium. The explanation, in this general shape, appears to reduce the phenomenon to its simplest interpretation; and the theory with regard to the composite structure of the chemical atoms provides a cause plainly adequate to produce effects of the kind which scientific experiments have revealed.

138. Bismuth is the most powerful of diamagnetic elements. Its density is 9·82, or nearly equal to that of silver, and one half that of gold. Its atomic number,

formerly rated at 71, is now often taken at three times the amount, or 213, while the specific heat would lead to a number half this amount, or 106½. If we assume it to consist, like iron, of four planes or cycles, each of them, according to the value assumed, will have 18, 27, or 54 units. We may easily, then, conceive its structure to be some slightly modified form of a flat cylinder. Now bismuth "in structure is highly lamellated. It is brittle when cold, but may be hammered into plates while warm. At 476° it fuses and sublimes in close vessels at a red heat. It is a less perfect conductor of heat than most other metals." All these features agree perfectly with the hypothesis of a flatness in the structure of the atoms. The metal is also wanting in ductility, which is another contrast to the properties of iron. On the other hand, its low conducting power for heat, and also for electricity, stands in diametrical opposition to the hypothesis which explains the diamagnetism of copper and silver by their conducting power. For the conductive power of bismuth is fifty times less than that of silver, and six times less than that of iron; while its atomic density, in the highest estimate, is more than half that of silver, and in the lower estimates, of 71, or 106½, is actually greater. So that, on the principles of the theory, its diamagnetism ought to be much less, instead of being far more powerful. But, on the present view, a lamellated structure and fusibility, or else a high electric order, should mark diamagnetics; and bismuth has the first two properties in a high degree.

139. Antimony is the next metal in diamagnetic order. Now this stands high in the electric series. It has also a large atomic number, and on both accounts must be conceived to have a radial and not an axial structure.

Again, it is "a brittle metal, fuses at 810°, and on cooling acquires a highly lamellated structure, and is volatile at an intense temperature" (Turner, p. 493). It is wanting both in ductility and tenacity. All these characters, as in the last metal, agree perfectly with the supposition of the flat form of its atoms, and of a cohesion by the edges rather than the flat sides or ends.

140. Zinc, the third diamagnetic metal, lies much lower than the two others, and especially than antimony and arsenic, in the voltaic scale. Its atomic number, also, is 32·5, or only slightly greater than that of iron, nickel, and cobalt, the magnetic metals. Its electro-positive character is very conspicuous, being constantly employed, in the electric circuit, for the positive element, and capable of very easy oxidation.

These properties are rather anomalous, and make it less easy to conjecture a form of the atom, which will satisfy all the conditions. But still other properties agree well with the theory. "Its texture is lamellated; at low or high degrees of heat it is brittle; but at temperatures between 210° and 500° it is malleable and ductile, which enables it to be rolled or hammered into sheets of considerable thinness." It may perhaps be a question that still requires to be settled, whether the diamagnetism of zinc belongs to it at all temperatures, or whether it is not a variable quality, capable of reversal, when the temperature is varied between the above limits.

141. Tin and cadmium are the two next diamagnetic metals. Their atoms 58 and 56 are nearly of the same value. In electric order, tin is just above and cadmium below the three magnetic metals. The two metals resemble each other. Cadmium is both ductile and malleable;

tin is malleable, but of much lower ductility. They both fuse at 442°. Tin "is soft and inelastic, and when bent emits a peculiar crackling noise." These properties all agree with the general conception of a lateral cohesion, and a lamellated character, reversing the characters of iron and platinum, or rendering the ductility of tin and cadmium inferior to their malleability.

142. Gold, silver, and copper come near together in electric order, and are also feebly diamagnetic. They are also ductile and tenacious, and form a sub-class quite distinct from the previous metals. Their properties all agree with the hypothesis that their breadth is slightly in excess of their length, either in extent or angularity, so as to admit only of a slight excess of polarity, either way, in a bar or cylinder of the metal, but to render the arrangement of the atoms more easy, when the equators are more numerous than the poles in the direction of the length.

143. Mercury, water, ether, and alcohol are all diamagnetic. These are the most characteristic liquids, mercury being the only fluid metal, and water the almost universal solvent. The fluidity of mercury, and the weight of its atom, imply that its axes and moments of rotation are very nearly equal. In this case the mere fact of rotation, as a liquid, will give a slight oblateness to the compound atom, and will therefore dispose it to arrange its equators, rather than its poles, in the line of greatest length. Again, water, for the same reason, must be conceived to have its atoms revolving around the axis of greatest, not of least moment. Hence its composition will be oblate and not prolate, and will be favourable to a disposition of the equators in the line of the length. The

question whether these substances are still diamagnetic when hard frozen deserves, perhaps, a series of careful experiments.

144. While the leading liquids, water, mercury, and alcohol are diamagnetic, oxygen is found to be strongly magnetic; but other gases, except air and some compounds of oxygen, are very nearly or altogether neutral. This property of oxygen gas agrees fully with the hypothesis. Its typical character, by the previous chapters, would seem to be a flat cycle in rapid rotation. Its gaseous expansion reverses the condition of a liquid, and enables its atoms to move freely among each other, and to receive and retain, under magnetic power, a partial parallelism with each other, so that the equators shall be in the line of the breadth, and the faces in that of the length; when in solids the consequence would be the reverse, from the want of room for the free adjustment of the poles, so that the need of compact arrangement would place the equators in the line of the length. The neutrality of other gases is less easy to explain. But if there is originally a near balance of polar tendency, the magnetic influence may be too weak to determine a fixed preponderance, since all the particles are moving constantly around each other, and thus tend to vary perpetually the direction of their poles.

CHAPTER XIII.

ON TERRESTRIAL PHYSICS IN GENERAL.

145. The doctrine that our present chemical elements are compound, and mutually convertible under favourable circumstances of extreme pressure or electric force, opens a wide field of inquiry with regard to several classes of phenomena, of which the causes have been hitherto unexplained.

All the varieties, which chemistry makes known to us, on the present view, are the results of special arrangements of centres of force; and are included between two extremes, the free ether of planetary space, and the strong compression of the central portions of the sun, the earth, and the other planets. Nature is thus like a tree of which the roots are hidden from sight, deep towards the earth's centre, while its lightest blossoms are in the height of the atmosphere, bordering on the ambient ether; and the stem, the branches, and the foliage, are the various minerals of the solid strata near the surface, and the liquid elements and gases of the sea, the land, and the air. Chemical composition is included between these limits, and probably disappears alike under the intense pressure near the centre of the earth, and in the diffused and exceedingly rare matter which remains still uncondensed throughout the solar system.

146. The Central Structure of the Earth is the first main question, which has given rise to much speculation. Its spheroidal shape has naturally led to the inference of its former fluidity. And since the temperature rises steadily, as we descend into the deepest mines, it has been a popular hypothesis that the whole mass of the earth, even now, is liquid with heat below the depth of thirty or forty miles. Of late this conclusion has been questioned on astronomical grounds, and it has been inferred, from the phenomena of nutation, that a solid shell, 800 or 1000 miles in thickness, is required to satisfy the mechanical conditions of this disturbance.

There is, however, a great oversight in the reasoning, which infers the present fluidity of the central mass of the earth from the rate of increase of temperature in mines. For the point of fusion must plainly rise with the increase of pressure, as the boiling-point is altered in liquids by the same cause. But while both the pressure and the sensible heat increase towards the centre, and have an opposite tendency to cause and to hinder fusion, it is far more natural to suppose that the pressure is the exciting cause of the heat, than that both are quite independent, or heat the cause of pressure. Now if the increase of temperature is really a secondary result from the increase of pressure, the natural consequence must be, that the parts next the centre, where the pressure is greatest, must be solid and not liquid; and that the liquefying power of the heat, if it exist anywhere, must be confined to parts much nearer to the surface, but too low for the solidifying power of the cooling radiation from the surface. Even in the most heated part it may well be doubted whether the normal state is one of actual fluidity, and not rather of constrained solidity under intense pressure, so as to produce instant fusion, or even partial evaporation, whenever, from any disturbing cause, that pressure is lessened or removed.

The Central Heat of the Earth, on the present view of matter and ether, will be only a result of the pressure due to its mutual gravity, and to the condensation, by which the attached ether is forced outwards and upwards from the parts nearest the centre. This extrusion of the ether will counteract the effects of pressure in the parts nearer to the surface, and may retain a considerable portion in a state of igneous fluidity. But the surface itself is cooled by the radiation of heat into space, and hence becomes a solid crust. The middle portion, whether actually fluid, or solid through the pressure, must liquefy or turn into vapour if the pressure be locally removed, and may thus give rise to all the phenomena of volcanic action.

147. *Mineral Veins.*

Some of the usual phenomena, very perplexing and unaccountable on the usual hypothesis, find an easy explanation on the present view. The metals are rarely found native in a metallic state, but in sulphurets, oxides, sulphates, carbonates, or more complex combinations. Some of them appear to have a special friendship and affinity, as silver is often found in ores of lead, and platinum, iridium, palladium are derived generally from the same ore. Also, cases are not unfrequent, where a vein producing one metal, on traversing a different stratum, produces another metal, copper replacing tin, or tin copper, and so in other combinations.

All these facts agree with the view that the different metals are formed, under strong pressure, deep below the surface of the earth; but that their state, as found in the veins, depends on the mechanical conditions under which the veins are formed, and on the character of the stratum or country through which they successively pass, while expanding and cooling in their escape from the original pressure.

The atomic numbers of silver and lead are 108 and 103·5, which differ little from each other. Lead is far more plentiful, but a small proportion of silver is found to be contained in the most common lead ores. Platinum and iridium have almost exactly the same atomic number, and are associated usually in the ore. Copper veins change into tin, or tin into copper, while the presence of iron is held to be an unfavourable sign, in searching for these metals. These facts, and many others of the same kind, are comparatively easy to understand, if the metals are not simple, but composite, and their formation depends on special circumstances of heat, pressure, and electric influence, so that a change in one of them may alter the nature of the metallic product.

148. *Chemical Relations of Geological Eras.*

A striking fact, in geology, is the special development of some chemical element or compound at particular eras of the earth's history.

To take a single example: the great development of carboniferous deposits at a certain period is generally supposed to imply simply an immense vegetation, receiving carbon from an atmosphere charged with carbonic gas. When, however, we consider the thickness of these seams of coal, and the vast surfaces over which they extend, it is difficult to conceive of an atmosphere which could furnish the supply required. On the other hand, since carbon, from the lightness of its atom, and the gaseous nature of its compounds with hydrogen, oxygen, and nitrogen, forms a transition between the earthy and metallic bases and the gaseous elements, it is natural to conceive that there must be some midway stage of geological change, favourable to its development, when vegetable life would be most amply supplied with it, and be the means of laying it up in store for after ages of mankind.

149. The saltness of the ocean is another leading fact, of which no explanation is offered on the principles of our actual chemistry. It has also been shewn that the saltness increases at greater depths. But if the atom of sodium, as has been shewn to be probable, bears a simple relation to the atoms of water, in that form which they would assume under strong pressure, and chlorine or muriatic acid has a relation to it of the opposite kind, so that one may result from water compressed axially, and the other from water expanded radially, under the opposite influence of intense pressure, we shall have a simple key to a phenomenon so fundamental in the scheme of nature, and which has baffled ordinary methods of explanation.

150. Another striking fact, of which no satisfactory account has yet been given, is the constitution of the atmosphere. It consists, by volume, of four parts of nitrogen and one of oxygen; or, if 8 and 14 are taken for the atomic weights, of two atoms of nitrogen and one of oxygen. This proportion is maintained wherever the atmosphere is analyzed, in spite of the perpetual change to which it is exposed from the action of animal and vege-

table life. On the other hand, it is a mixture, and not a chemical compound, though this exact chemical relation between its two elements is still maintained.

If, however, we accept the first principle of the present theory, that all the elements are convertible under proper conditions of pressure or electricity, a relation of singular simplicity reveals itself. Two volumes of hydrogen, and one of oxygen, united by the electric spark, produce two volumes of aqueous vapour, and thus, from synthesis and analysis at the surface of the earth, water is viewed as the compound of these two gases in this proportion. Tried, however, by the test of volumes, it is one atom of hydrogen, and half an atom of oxygen gas, which form an atom of water. But four atoms of nitrogen gas, and one of oxygen gas, the ratio in the atmosphere, give 72 for their atomic number, which is precisely the atomic weight of eight atoms of water. Now as aqueous vapour mounts higher in the air, it must plainly tend to a more gaseous condition. And hence, as in the case of forced vibrations, or synchronous movements in general, the atoms of nitrogen and oxygen must have a certain tendency to assimilate this vapour to themselves; while the relation of the numbers is such as to ensure that its decomposition into these two gases, if capable of being effected at all, will be only in this proportion. The case will be similar to the relation between the periodic times of satellites, where a near approach, by a constant cause, is turned into an exact proportion. Supposing an approach to this ratio at first, and a power, in the highest region of the air, to resolve aqueous vapour into these two gases, a cause perpetually at work would tend to secure finally this exact proportion throughout the whole.

Such a tendency of the oxygen and nitrogen of the atmosphere to transform aqueous vapour, when extremely rarefied, must result at once from the present theory. For the atoms of water, whether arranged in three triplets, or in two fours and a unit, must be changed by the absorption of latent heat, in ascending, till they take the form of a single cycle of nine. But when this cycle is expanded, so that the cohesion of its nearest units is weakened, its rotation will be out of harmony with those of the oxygen and nitrogen in their nearest appulse. If four pairs of them, by the rotatory rhythm of the nitrogen atoms, are turned into nitrogen, the four quaternions separated from them will equally combine to form an atom of oxygen. The change, if it occurs at all, must take place at the upper limit of the presence of aqueous vapour, and consequently above the region of the highest clouds, where the exciting power of the solar rays must be the greatest, and the cohesion of the cycles is reduced much below its amount at the surface of the earth. It is therefore no objection to the view, that water cannot be separated into these two gases under the conditions which actually obtain at the surface of the earth.

151. The origin and composition of meteoric stones has given rise to much speculation. The general opinion, at present, is that they are wholly cosmical, or extra-terrestrial, especially since the occurrence of shooting-stars is, to a great extent, periodic, and occurs when the earth is in one or two special parts of her annual orbit. But two serious difficulties still remain. First, it seems clear that, in many instances, meteoric stones or thunderbolts, are directly the result of thunderstorms, or violent electric disturbance within the atmosphere, especially in tropical

countries. And, next, it seems strange and unaccountable that floating matter in planetary space should consist mainly of two or three metals, to the exclusion of others, and these metals the same which are distinctively magnetic. If our present elements are viewed as ultimate, independent forms of matter, no key is supplied to either of these phenomena.

Now let us compare the atomic weights of these meteoric and magnetic metals with those of the atmospheric gases, and a relation of singular simplicity discloses itself. Nitrogen is the chief constituent of the atmosphere, iron the most magnetic metal, and the most abundant constituent of meteoric stones, some of which are indeed masses of meteoric iron. The atomic weight of nitrogen is 14, of iron 28; or two atoms of nitrogen, by weight, will exactly form one atom of iron. Again, the atoms of nickel cobalt are each reckoned at 29½. But as two atoms of nitrogen would be 28, so an atom of nitrogen and one of oxygen would make 30, almost exactly the atomic weight of each of these metals. The magnetic character, also, of these elements in meteoric stones is a clear sign that they result, in some way, from a powerful electric influence.

Again, this view of the formation of meteoric iron, and the kindred metals, carries out a plain analogy. In common thunder showers, the electric action is moderate, and its effect is to condense aqueous vapour into large or small drops of rain. In violent thunder storms, the electric force is greater, and the vapour is not only condensed into water, but compacted into solid hail stones, sometimes of uncommon size. In thunder storms still more violent, or in disturbances caused by foreign matter encountering the earth's annual motion, it is quite consistent to believe that the electric action would be more powerful still, and might condense the atoms of the air itself, crushing out the ether which separates the nearest, and converting them, by pairs, into the atoms of solid and magnetic metals.

On the view here proposed, the nature of such a transformation of nitrogen into iron may be easily conceived. Intense ethereal pressure, from without, will first bring two cycles of fourteen axially together. The same pressure, acting on their circumference, will force them to resolve themselves each into two cycles of seven; and the same disturbance, or ejection of the central ether, will force one of the seven units in each cycle to pass inward, and occupy the centre. The resistance to further compression will then be completed, while intense heat will of course be developed by this immense condensation, and the closer approach of the component units of matter.

152. "The atmospheric hypothesis," Dr Lardner remarks, "is subject to objections so unanswerable, that it may be considered as altogether set aside. In order to suppose it probable that aerolites can be formed in the atmosphere, we must shew that their constituent elements can exist there. But the most rigorous analysis has never detected in the air any of the constituents of meteoric stones...The actual velocities with which they are known to strike the earth could never be acquired under the mere agency of terrestrial gravity. If the velocity of the meteorites be incompatible with this theory, their direction is still more, and their obliquity could never be produced by any atmospheric current" (*Mus.* I. p. 138).

These objections, it is plain, are wholly inapplicable to the view here proposed. The materials of the meteoric metals will be present abundantly, being the atmospheric gases themselves. The same intense electric energy which is required for their formation will be clearly adequate to impress upon them an oblique direction, and an immense velocity. The explanation includes equally the phenomena which indicate a cosmical, and those which point to a terrestrial origin. In the former case, the uncondensed matter of outer space, encountering the earth in its orbit, may be condensed, and condense the air, in passing through its highest and most electric portions; while, in other cases, violent electric energy within the atmosphere itself may have the like effect without foreign collision. The periodical shooting stars will belong to one class, and the meteors of thunder storms, or those which are proved to be of moderate height, and in some instances to move upward, will be of the other.

"In general, the trains of shooting stars have the same hollow, cylindrical appearance as the tails of comets, their inner part appearing to be void of luminous matter, and a further resemblance to comets is seen in the curved form which they sometimes assume. Beccaria and Vassali considered them to be lines of electric sparks, an hypothesis which has been abandoned. Lavoisier, Volta, and others, supposed that hydrogen, accumulated by its lightness in the higher regions of the atmosphere, was inflamed. But the law of gases which gives them a tendency to mingle, notwithstanding their specific gravity, sets aside this hypothesis...Poisson affirmed the probability of an atmosphere of electricity surrounding the earth, and higher than the atmosphere of air, and that the meteorite rushing through this atmosphere would decompose this electric fluid, like the friction of a machine."

All these divergent suggestions are harmonized and

reconciled by the results of the present theory. The outmost strata of the atmosphere, where chemical structure and gaseous repulsion cease, must be composed of igneous matter, or hydrogen gas, saturated with ether by the remoteness of its atoms from each other, and their diminished motion. There will thus be an atmosphere of positive electricity, with a shell of hydrogen for the material vehicle on which it depends. The phenomena of shooting stars will depend on electric disturbance, or the swift passage of foreign matter, in this highest region, where the exceeding rarity of the matter makes the results to be electrical or luminous alone. The resemblance of the trains to the tails of comets will result at once from the sameness of the conditions. But when the electric disturbance, or the rapid transit, takes place in lower regions, where the gases of the atmosphere subsist unchanged, the effect of the intense compression may be the formation of meteoric iron and the other kindred metals, joined also, in some cases, with the electro-negative elements.

"The meteorites which fell at Agram in India, those which were found at Sisim, in the Jeniseisk government, and those brought by Humboldt from Mexico, contained as much as 96 per cent. of very malleable iron." Can we suppose it in the least probable, that the floating matter of the planetary spaces, which our earth encounters in its annual orbit, consists of this one magnetic metal, ready formed, in so large a proportion?

153. *Connexion between the Sun and Terrestrial Magnetism.*

It has been ascertained, within the last few years, that there is an intimate relation between the solar light and the oscillations or variations of the earth's magnetism.

The observations of Schwabe seem to prove that the maximum and minimum of the solar spots, in a decennial period, correspond with a maximum and minimum of magnetic power on the surface of the earth.

On the origin of the period itself it would perhaps be premature to offer any conjecture. But the present theory implies at once an intimate relation between the solar light and terrestrial magnetism. For magnetism, on this view, is only a special result, in bodies of a peculiar atomic constitution, of the general fact of polarity, arising from moments of rotation, and the tendency of similar motions, through the intense elasticity of free ether, to generate each other, and of dissimilar motions to interfere with each other. Hence it seems to follow, as a natural consequence, not only that the magnetic power of the earth depends upon, and is derived from, its own rotation on its axis; but that the same cause, in the greater mass and with the swifter equatorial motion of the sun, must produce a magnetism far more intense, and capable probably of a sensible effect, even at his immense distance. And since the vehicle of its transmission must be the luminous atmosphere, any variation, such as that of the solar spots, implying a periodic change in that atmosphere, may be expected, on the principles of the theory, to have some effect, and probably a sensible effect, on the variations of terrestrial magnetic power.

154. *Source of the Solar Light and Heat.*

Another subject, which has given birth of late to some ingenious speculation, is the means by which the supply of the sun's light and heat is sustained. The latest hypothesis, recommended rather by the difficulties which encompass every other, than by its own elegance, is that the surface of the sun is hammered into a white heat by the continual impact of revolving bodies, or extra-solar planetoids, since heat and mechanical force are strictly convertible. Combustion, it is thought, could not sustain the immense efflux of light and heat without causing a sensible waste of the sun's body; so that the impact hypothesis, however strange and unpoetical, is conceived to be alone capable of meeting the conditions of the physical problem.

The present theory suggests a very different explanation, which accords far better with all the analogies of nature. The mechanical energy which the sun exerts, in the light and heat which flow from it to all the planets, must plainly, on this view of matter and ether, be a direct result from the rotation of the immense central mass. It cannot, then, be the same in all directions. By virtue of its revolution the sun must form a vast magnet, and a special repulsion, propagated through the ether in the plane of the equator, be compensated by a special attraction in the line of the poles. The mechanical energies which are thrown out continually in light and heat cannot be wasted, or lose themselves in space, and must return, by a cosmical circuit, to their natural home. The ethereal force will thus diffuse itself in swift undulations to the verge of the solar system; and travelling at right angles, when this impulse ceases or grows feeble, will return to the centre in the line of the solar axis.

This view seems to meet with partial, indirect confirmation, in a singular astronomical fact. The orbits of the planets and their satellites, in general, lie near to the plane of the ecliptic. But the satellites of Uranus, the outmost planet except Neptune, are a remarkable exception. They are inclined at an angle of 78°, and since their motion is retrograde, 102° may rather be viewed as their true inclination. All recent discovery also tends to prove the existence of strong directive forces, whatever be their mechanical cause, on which electric currents and magnetic action depend. We may infer that the directive force, at the distance of Uranus, is no longer in the ecliptic plane, but more nearly at right angles. We should thus have reached, or nearly, the middle stage in a grand cosmical circuit, where the ethereal energy, emanating from the sun as a centrifugal force in the plane of the equator, begins to travel at right angles, that it may return at length, in the line of the axis, to the centre again. It is even conceivable that in this way magnetic relations of polar attraction and repulsion may be generated between distant solar or sidereal systems on the largest scale.

There are many relations among the known facts of science which acquire a new significance, when compared with the present theory of the constitution of matter, and may perhaps open new fields of inquiry or future discovery. But in the present stage of its development, it may be wiser to abstain from further suggestions, which must partake largely of a tentative and conjectural character. The main principle of the theory, the inductive basis on which it rests, and the simple solution it supplies to all those classes of phenomena which form the great branches of physical science, have now been unfolded at sufficient length for a first Essay. It is now committed to the candid reflection of men of science, with the strong confidence, that it either supplies the true and sufficient key to the remaining secrets of inorganic nature, or at least will assist other and more skilful labourers in its discovery, whenever it shall please the Supreme Architect to make a second and fuller revelation of those laws, which preside over all the countless phenomena of change in the outward world.

CAMBRIDGE: PRINTED BY C. J. CLAY, M.A. AT THE UNIVERSITY PRESS.

MACMILLAN & CO.'S

LIST OF

NEW AND POPULAR WORKS.

SECOND EDITION.

RAVENSHOE.

BY HENRY KINGSLEY, Author of 'Geoffry Hamlyn.'

3 Vols. crown 8vo. cloth, £1 11s. 6d.

'Admirable descriptions, which place "Ravenshoe" almost in the first rank of novels. Of the story itself it would really be difficult to speak too highly. The author seems to possess every essential for a writer of fiction.'—London Review.

SECOND EDITION.

RECOLLECTIONS OF GEOFFRY HAMLYN.

BY HENRY KINGSLEY. Crown 8vo. cloth, 6s.

'Mr. Henry Kingsley has written a work that keeps up its interest from the first page to the last—it is full of vigorous stirring life. The descriptions of Australian life in the early colonial days are marked by an unmistakable touch of reality and personal experience. A book which the public will be more inclined to read than to criticise, and we commend them to each other.'—Athenæum.

SECOND EDITION.

TOM BROWN AT OXFORD.

3 Vols. £1 11s. 6d.

'A book that will live. In no other work that we can call to mind are the finer qualities of the English gentleman more happily portrayed.'—Daily News.

'The extracts we have given can give no adequate expression to the literary vividness and noble ethical atmosphere which pervade the whole book.'—Spectator.

TWENTY-EIGHTH THOUSAND.

TOM BROWN'S SCHOOL DAYS.

BY AN OLD BOY. Fcp. 8vo. 5s.

'A book which every father might well wish to see in the hands of his son.'—Times.

EIGHTH THOUSAND.

SCOURING OF THE WHITE HORSE.

By the author of 'Tom's Brown's School Days.'

With numerous Illustrations by Richard Doyle. Imperial 16mo. Printed on toned paper, gilt leaves, 8s. 6d.

'The execution is excellent. . . . . Like "Tom Brown's School Days," the "White Horse" gives the reader a feeling of gratitude and personal esteem towards the author. The author could not have a better style, nor a better temper, nor a more excellent artist than Mr. Doyle to adorn his book.'—Saturday Review.

WORKS BY THE REV. CHARLES KINGSLEY.

CHAPLAIN IN ORDINARY TO THE QUEEN, RECTOR OF EVERSLEY,
AND PROFESSOR OF MODERN HISTORY IN THE UNIVERSITY OF CAMBRIDGE.

WESTWARD HO!

NEW AND CHEAPER EDITION. Crown 8vo. cloth, 6s.

'Mr. Kingsley has selected a good subject, and has written a good novel to an excellent purpose.'—TIMES.

'We thank Mr. Kingsley heartily for almost the best historical novel, to our mind, of the day.'
FRASER'S MAGAZINE.

TWO YEARS AGO.

NEW AND CHEAPER EDITION. Crown 8vo. cloth, 6s.

'In "Two Years Ago," Mr. Kingsley is, as always, genial, large-hearted, and humorous; with a quick eye and a keen relish alike for what is beautiful in nature and for what is genuine, strong, and earnest in man.'—GUARDIAN.

ALTON LOCKE, TAILOR AND POET.

NEW EDITION. Crown 8vo. cloth, 4s. 6d. With New Preface.

☞ This Edition is printed in crown 8vo. uniform with 'Westward Ho!' &c., and contains a New Preface.

THE HEROES.

GREEK FAIRY TALES FOR THE YOUNG.

SECOND EDITION, with Illustrations. Royal 16mo. cloth, 5s.

'A charming book, adapted in style and manner, as a man of genius only could have adapted it, to the believing imagination, susceptible spirit of youths.'—BRITISH QUARTERLY REVIEW.

'Rarely have these heroes of Greek tradition been celebrated in a bolder or more stirring strain.'—SATURDAY REVIEW.

ALEXANDRIA AND HER SCHOOLS.

Crown 8vo. cloth, 5s.

THE LIMITS OF EXACT SCIENCE

AS APPLIED TO HISTORY.

INAUGURAL LECTURE AT CAMBRIDGE. Crown 8vo. 2s.

PHAETHON:

LOOSE THOUGHTS FOR LOOSE THINKERS.

THIRD EDITION. Crown 8vo. 2s.

NEW AND CHEAPER EDITION (FIFTH THOUSAND).

Handsomely printed on toned paper and bound in extra cloth. With Vignette and Frontispiece from Designs by the author. Engraved on Steel by C. H. Jeens. 4s. 6d.

THE LADY OF LA GARAYE.

By the Hon. Mrs. NORTON. Dedicated to the Marquis of Lansdowne.

'The poem is a pure, tender, touching tale of pain, sorrow, love, duty, piety, and death.'
EDINBURGH REVIEW.

'A true poem, noble in subject and aim, natural in flow, worthy in expression, with the common soul of humanity throbbing in every page through wholesome words.'—EXAMINER.

GOBLIN MARKET AND OTHER POEMS.

By CHRISTINA G. ROSSETTI. With Two Illustrations from Designs by D. G. Rossetti. Fcp. 8vo. cloth, 5s.

'As faultless in expression, as picturesque in effect, and as high in purity of tone, as any modern poem that can be named.'—SATURDAY REVIEW.

THE LUGGIE AND OTHER POEMS.

By DAVID GRAY. With a Preface by R. MONCKTON MILNES, M.P. Fcp. 8vo. cloth, 5s.

'Full of the purest and simplest poetry.'—SPECTATOR.

SECOND EDITION.

EDWIN OF DEIRA.

By ALEXANDER SMITH. Fcp. 8vo. cloth, 5s.

'The poem bears in every page evidence of genius controlled, purified, and disciplined, but ever present.'—STANDARD.

'A felicitous and noble composition.'—NONCONFORMIST.

BY THE SAME AUTHOR.

A LIFE DRAMA AND OTHER POEMS.

4th Edition, 2s. 6d.

CITY POEMS.

5s.

BLANCHE LISLE AND OTHER POEMS.

By CECIL HOME. Fcp. 8vo. cloth, 4s. 6d.

'The writer has music and meaning in his lines and stanzas, which, in the selection of diction and gracefulness of cadence, have seldom been excelled.'—LEADER.

'Full of a true poet's imagination.'—JOHN BULL.

THE POEMS OF ARTHUR HUGH CLOUGH,

SOMETIME FELLOW OF ORIEL COLLEGE, OXFORD.

With a Memoir by F. T. PALGRAVE. Fcp. 8vo. cloth, 6s.

'Few, if any, literary men of larger, deeper, and more massive mind have lived in this generation than the author of these few poems, and of this the volume before us bears ample evidence...... There is nothing in it that is not in some sense rich either in thought or beauty, or both.'—SPECTATOR.

Uniform with 'WESTWARD HO!' 'GEOFFRY HAMLYN,' &c.

THE MOOR COTTAGE:

A TALE OF HOME LIFE.

By MAY BEVERLEY, author of 'Little Estella, and other Fairy Tales for the Young.' Crown 8vo. cloth, price 6s.

'This charming tale is told with such excellent art, that it reads like an episode from real life.'—ATLAS.

'The whole plot of the story is conceived and executed in an admirable manner: a work which, when once taken up, it is difficult to put down.'—JOHN BULL.

Uniform with 'WESTWARD HO!' 'GEOFFRY HAMLYN,' &c.

ARTIST AND CRAFTSMAN.

Crown 8vo. cloth, price 6s.

'Its power is unquestionable, its felicity of expression great, its plot fresh, and its characters very natural... Wherever read, it will be enthusiastically admired and cherished.'

MORNING HERALD.

Uniform with 'WESTWARD HO!' 'GEOFFRY HAMLYN,' &c.

A LADY IN HER OWN RIGHT.

By WESTLAND MARSTON. Crown 8vo. cloth, price 6s.

'Since "The Mill on the Floss" was noticed, we have read no work of fiction which we can so heartily recommend to our readers as "A Lady in her Own Right:" the plot, incidents, and characters are all good: the style is simple and graceful: it abounds in thoughts judiciously introduced and well expressed, and throughout a kind, liberal, and gentle spirit.'

CHURCH OF ENGLAND MONTHLY REVIEW.

THE BROKEN TROTH:

A TALE OF TUSCAN LIFE FROM THE ITALIAN.

By PHILIP IRETON. 2 vols. fcp. 8vo. cloth, 12s.

'The style is so easy and natural. . . The story is well told from beginning to end.'—PRESS.

'A genuine Italian tale—a true picture of the Tuscan peasant population, with all their virtues, faults, weaknesses, follies, and even vices. . . The best Italian tale that has been published since the appearance of the 'Promessi Sposi' of Manzoni. . . The 'Broken Troth' is one of those that cannot be read but with pleasure.'—LONDON REVIEW.

GARIBALDI AT CAPRERA.

By COLONEL VECCHJ. With Preface by Mrs. Gaskell, and a View of Caprera. Fcp. 8vo. cloth, 3s. 6d.

'After all has been told, there was something wanting to the full and true impression of the Patriot's character and mode of life; as every one who reads this artless and enthusiastic narration will certainly admit. Mrs. GASKELL says she knows that "every particular" of this full and minute account may be relied upon; and it has an air of truth that commends it even when it is most extravagant in its admiration."—NONCONFORMIST.

ROME IN 1860.

By EDWARD DICEY, author of 'Life of Cavour.' Crown 8vo. cloth, 6s. 6d.

'So striking and apparently so faithful a portrait. It is the Rome of *real* life he has depicted.'—SPECTATOR.

THE ITALIAN WAR OF 1848-9,

And the last Italian Poet. By the late HENRY LUSHINGTON, Chief Secretary to the Government of Malta. With a Biographical Preface by G. STOVIN VENABLES. Crown 8vo. cloth, 6s. 6d.

'Perhaps the most difficult of all literary tasks—the task of giving historical unity, dignity, and interest to events so recent as to be still encumbered with all the details with which newspapers invest them—has never been more successfully discharged. . . Mr. Lushington, in a very short compass, shows the true nature and sequence of the event, and gives to the whole story of the struggle and defeat of Italy a degree of unity and dramatic interest which not one newspaper reader in ten thousand ever supposed it to possess.'—SATURDAY REVIEW.

EARLY EGYPTIAN HISTORY.

FOR THE YOUNG.

WITH DESCRIPTIONS OF THE TOMBS AND MONUMENTS.

By the Author of 'Sidney Grey,' &c. and her Sister. Fcp. 8vo. cloth, 5s.

'Full of information without being dull, and full of humour without being frivolous; stating in the most popular form the main results of modern research. . . . We have said enough to take our readers to the book itself, where they will learn more of Ancient Egypt than in any other popular work on the subject.'—LONDON REVIEW.

DAYS OF OLD;

OR, STORIES FROM OLD ENGLISH HISTORY.

FOR THE YOUNG.

By the Author of 'Ruth and Her Friends.' With a Frontispiece by W. HOLMAN HUNT. Royal 16mo. beautifully printed on toned paper and bound in extra cloth, 5s.

'A delightful little book, full of interest and instruction. . . fine feeling, dramatic weight, and descriptive power in the stories. . . They are valuable as throwing a good deal of light upon English history, bringing rapidly out the manners and customs, the social and political conditions of our British and Anglo-Saxon ancestors, and the moral always of a pure and noble kind.'—LITERARY GAZETTE.

HOW TO WIN OUR WORKERS.

A Short Account of the Leeds Sewing School for Factory Girls. By Mrs. HYDE. Dedicated by permission to the EARL of CARLISLE. Fcp. 8vo. limp cloth, 1s. 6d.

This work is intended to exhibit the successful working of an Institution for bringing the Working-girls of a large town into communication and sympathy with those who are separated from them by social position.

'A little book brimful of good sense and good feeling.'—GLOBE.

OUR YEAR.

Child's Book in Prose and Rhyme. By the author of 'John Halifax.' With numerous Illustrations by CLARENCE DOBELL. Royal 16mo. cloth, gilt leaves, 5s.

'Just the book we could wish to see in the hands of every child.'—ENGLISH CHURCHMAN.

LITTLE ESTELLA,

AND OTHER FAIRY TALES.

By MAY BEVERLEY. With Frontispiece. Royal 16mo. cloth, gilt leaves, 5s.

'Very pretty, pure in conception, and simply, gracefully related . . . genuine story-telling.'
DAILY NEWS.

MY FIRST JOURNAL.

A Book for Children. By GEORGIANA M. CRAIK, author of 'Lost and Won.' With Frontispiece. Royal 16mo. cloth, gilt leaves, 4s. 6d.

'True to Nature and to a fine kind of nature. . . . The style is simple and graceful. As a work of Art, clever and healthy-toned.'—GLOBE.

AGNES HOPETOUN'S SCHOOLS AND HOLIDAYS.

By Mrs. OLIPHANT, author of 'Margaret Maitland.' With Frontispiece Royal 16mo. cloth, gilt leaves, 5s.

'Described with exquisite reality . . . teaching the young pure and good lessons.'
JOHN BULL.

DAYS OF OLD:

STORIES FROM OLD ENGLISH HISTORY.

By the author of 'Ruth and Her Friends.' With Frontispiece. Royal 16mo. cloth, gilt leaves, 5s.

'A delightful little book, full of interest and instruction . . fine feeling, dramatic weight, and descriptive power in the stories.'—Literary Gazette.

DAVID, KING OF ISRAEL.

A History for the Young. By JOSIAH WRIGHT, Head Master of Sutton Coldfield Grammar School. With Illustrations. Royal 16mo. cloth, gilt leaves, 5s.

'An excellent book . . well conceived, and well worked out.'—Literary Churchman.

RUTH AND HER FRIENDS.

A Story for Girls. With Frontispiece. Third Edition. Royal 16mo. cloth, gilt leaves, 5s.

'A book which girls will read with avidity, and cannot fail to profit by.'
Literary Churchman.

SECOND EDITION.

GEORGE BRIMLEY'S ESSAYS.

Edited by WILLIAM GEORGE CLARK, M.A. Public Orator in the University of Cambridge. With Portrait. Crown, 8vo. cloth, 5s.

CONTENTS:

I. TENNYSON'S POEMS.
II. WORDSWORTH'S POEMS.
III. POETRY AND CRITICISM.
IV. ANGEL IN THE HOUSE.
V. CARLYLE'S LIFE OF STERLING.
VI. ESMOND.
VII. MY NOVEL.
VIII. BLEAK HOUSE.
IX. WESTWARD HO!
X. WILSON'S NOCTES.
XI. COMTE'S POSITIVE PHILOSOPHY.

'One of the most delightful and precious volumes of criticism that has appeared in these days. . . To every cultivated reader they will disclose the wonderful clearness of perception, the delicacy of feeling, the pure taste, and the remarkably firm and decisive judgment which are the characteristics of all Mr. Brimley's writings on subjects that really penetrated and fully possessed his nature.' Nonconformist.

WORKS BY DAVID MASSON, M.A.

PROFESSOR OF ENGLISH LITERATURE IN UNIVERSITY COLLEGE, LONDON.

LIFE OF JOHN MILTON.

Narrated in connexion with the Political, Ecclesiastical, and Literary History of his time. Vol. 1. 8vo. with Portraits, 18s.

'Mr. Masson's Life of Milton has many sterling merits . . . his industry is immense; his zeal unflagging; his special knowledge of Milton's life and times extraordinary. . . . With a zeal and industry which we cannot sufficiently commend, he has not only availed himself of the biographical stores collected by his predecessors, but imparted to them an aspect of novelty by his skilful re-arrangement.'—EDINBURGH REVIEW.

BRITISH NOVELISTS AND THEIR STYLES;

Being a critical sketch of the History of British Prose Fiction. Crown 8vo. cloth, 7s. 6d.

'A work eminently calculated to win popularity, both by the soundness of its doctrine and the skill of its art.'—THE PRESS.

ESSAYS, BIOGRAPHICAL AND CRITICAL.

Chiefly on English Poets. By DAVID MASSON. 8vo. cloth, 12s. 6d.

CONTENTS:

'Mr. Masson has succeeded in producing a series of criticisms in relation to creative literature which are satisfactory as well as subtile—which are not only ingenious, but which possess the rarer recommendation of being usually just.'—THE TIMES.

RELIGIO CHEMICI.

By GEORGE WILSON, M.D. late Regius Professor of Technology in the University of Edinburgh. Crown 8vo. cloth.

With a Vignette Title Page by NOEL PATON, engraved by C. JEENS. Price 8s. 6d.

THE FIVE GATEWAYS OF KNOWLEDGE.

A popular work on the Five Senses. By GEORGE WILSON, M.D. Eighth Thousand. In fcp. 8vo. cloth, with gilt leaves, 2s. 6d. People's Edition in ornamental stiff cover, 1s.

THE PROGRESS OF THE TELEGRAPH.

By GEORGE WILSON, M.D. Fcp. 8vo. 1s.

MEMOIR OF GEORGE WILSON, M.D. F.R.S.E.

Regius Professor of Technology in the University of Edinburgh. By his Sister, JESSIE AITKEN WILSON. With Portrait. 8vo. cloth, price 14s.

'His life was so pregnant in meaning, so rich in noble deeds, so full of that spiritual vitality which serves to quicken life in others; it bore witness to so many principles which we can only fully understand when we see them in action: it presented so many real pictures of dauntless courage and of Christian heroism, that we welcome gratefully the attempt to reproduce it which has resulted in the volume before us. Miss Wilson has entered lovingly upon her task, and has accomplished it well.'—PRESS.

MEMOIR OF EDWARD FORBES, F.R.S.

Late Regius Professor of Natural History in the University of Edinburgh.

By GEORGE WILSON, M.D. F.R.S.E. and ARCHIBALD GEIKIE, F.R.S.E. F.G.S. of the Geological Survey of Great Britain. 8vo. cloth, with Portrait, 14s.

'We welcome this volume as a graceful tribute to the memory of as gifted, tender, generous a soul as Science has ever reared, and prematurely lost.'—LITERARY GAZETTE.

'It is long since a better memoir than this, as regards either subject or handling, has come under our notice. . . The first nine chapters retain all the charming grace of style which marked everything of Wilson's, and the author of the latter two-thirds of the memoir deserves very high praise for the skill he has used, and the kindly spirit he has shown. From the first page to the last, the book claims careful reading, as being a full but not overcrowded rehearsal of a most instructive life, and the true picture of a mind that was rare in strength and beauty.'—EXAMINER.

MEMOIR OF THE

LIFE OF THE REV. ROBERT STORY,

LATE MINISTER OF ROSNEATH, DUMBARTONSHIRE.

By ROBERT HERBERT STORY, Minister of Rosneath. Crown 8vo. cloth, with Portrait, 7s. 6d.

*** This volume includes several important passages of Scottish Religious and Ecclesiastical History during the Second Quarter of the present Century. Among others, the Row Controversy, the Rise of the Irvingite Movement, the Early History of the Free Church, &c. &c.

THE PRISON CHAPLAIN:

A MEMOIR OF THE REV. JOHN CLAY,

LATE CHAPLAIN OF PRESTON GAOL.

With selections from his Correspondence and a Sketch of Prison Discipline in England. By his SON. With Portrait, 8vo. cloth, 15s.

'It presents a vigorous account of the Penal system in England in past times, and in our own. . . It exhibits in detail the career of one of our latest prison reformers; alleged, we believe with truth, to have been one of the most successful, and certainly in his judgments and opinions one of the most cautious and reasonable, as well as one of the most ardent.'

Saturday Review.

MEMOIR OF GEORGE WAGNER,

LATE INCUMBENT OF ST. STEPHEN'S, BRIGHTON.

By JOHN NASSAU SIMPKINSON, M.A. Rector of Brington, Northampton. Third and cheaper Edition. Fcp. 8vo. 5s.

'A more edifying biography we have rarely met with . . . If any parish priest, discouraged by what he may consider an unpromising aspect of the time, should be losing heart . . . we recommend him to procure this edifying memoir, to study it well, to set the example of the holy man who is the subject of it before him in all its length and breadth, and then he will appreciate what can be done even by one earnest man; and gathering fresh inspiration, he will chide himself for all previous discontent, and address himself with stronger purpose than ever to the lowly works and lofty aims of the ministry entrusted to his charge.'

Literary Churchman.

FAMILY PRAYERS.

By the Rev. GEORGE BUTLER, M.A. Vice-Principal of Cheltenham College, and late Fellow of Exeter College, Oxford. Crown 8vo. cloth, 5s.

CAMBRIDGE CLASS BOOKS

FOR COLLEGES AND SCHOOLS

PUBLISHED BY

MACMILLAN & CO.

A set of Macmillan & Co.'s Class Books was exhibited in the Educational Department (Class 29) of the International Exhibition, and for which a Medal was awarded.

Arithmetic. For the use of Schools. By BARNARD SMITH, M.A., Fellow of St. Peter's College, Cambridge. New Edition. Crown 8vo. cloth, 4s. 6d.

A Key to the Arithmetic for Schools. By BARNARD SMITH, M.A., Fellow of St. Peter's College, Cambridge. Second Edition. Crown 8vo. cloth, 8s. 6d.

Arithmetic and Algebra, in their Principles and Application: with numerous systematically arranged Examples, taken from the Cambridge Examination Papers. By BARNARD SMITH, M.A. Fellow of St. Peter's College, Cambridge. Eighth Edition. Crown 8vo. cloth, 10s. 6d.

Exercises in Arithmetic. By BARNARD SMITH. With Answers. Crown 8vo. limp cloth, 2s. 6d. Or sold separately, as follows:— Part I. 1s. Part II. 1s. Answers 6d.

An Elementary Treatise on the Theory of Equations, with a Collection of Examples. By I. TODHUNTER, M.A. Fellow and Mathematical Lecturer of St. John's College, Cambridge. Crown 8vo. cloth, 7s. 6d.

Euclid. For Colleges and Schools. By I. TODHUNTER, M.A., Fellow and Principal Mathematical Lecturer of St. John's College, Cambridge. Pot 8vo. [In the Press.

Algebra. For the use of Colleges and Schools. By I. TODHUNTER, M.A. Fellow of St. John's College, Cambridge. Second Edition. Crown 8vo. cloth, 7s. 6d.

Plane Trigonometry. For Colleges and Schools. By I. TODHUNTER, M.A. Fellow of St. John's College, Cambridge. Second Edition. Crown 8vo. cloth, 5s.

A Treatise on Spherical Trigonometry. For the use of Colleges and Schools. By I. TODHUNTER, M.A. Fellow of St. John's College, Cambridge. Crown 8vo. cloth, 4s. 6d.

Examples of Analytical Geometry of Three Dimensions. By I. TODHUNTER, M.A. Fellow of St. John's College, Cambridge. Crown 8vo. cloth, 4s.

A Treatise on the Differential Calculus. With numerous Examples. By I. TODHUNTER, M.A. Fellow and Assistant Tutor of St. John's College, Cambridge. Third Edition. Crown 8vo. cloth, 10s. 6d.

A Treatise on the Integral Calculus. With numerous Examples. By I. TODHUNTER, M.A. Fellow and Assistant Tutor of St. John's College, Cambridge. Second Edition. Crown 8vo. cloth, 10s. 6d.

A Treatise on Analytical Statics. With numerous Examples. By I. TODHUNTER, M.A. Fellow of St. John's College, Cambridge. Second Edition. Crown 8vo. cloth, 10s. 6d.

First Book of Algebra. For Schools. By J. C. W. ELLIS, M.A., and P. M. CLARK, M.A. Sidney Sussex College, Cambridge. [Preparing.

Arithmetic in Theory and Practice. For Advanced Pupils. By J. BROOK SMITH, M.A. Part First. 164 pp. (1860). Crown 8vo. 3s. 6d.

A Short Manual of Arithmetic. By C. W. UNDERWOOD, M.A. 96 pp. (1860). Fcp. 8vo. 2s. 6d.

Introduction to Plane Trigonometry. For the use of Schools. By J. C. SNOWBALL, M.A. Second Edition (1847). 8vo. 5s.

Plane and Spherical Trigonometry. With the Construction and Use of Tables of Logarithms. By J. C. SNOWBALL, M.A. Ninth Edition, 240 pp. (1857). Crown 8vo. 7s. 6d.

Plane Trigonometry. With a numerous Collection of Examples. By R. D. BEASLEY, M.A. 106 pp. (1858). Crown 8vo. 3s. 6d.

Elementary Treatise on Mechanics. With a Collection of Examples. By S. PARKINSON, B.D. Second Edition, 345 pp. (1860). Crown 8vo. 9s. 6d.

A Treatise on Optics. By S. PARKINSON, B.D. 304 pp. (1859). Crown 8vo. 10s. 6d.

Elementary Hydrostatics. With numerous Examples and Solutions. By J. B. PHEAR, M.A. Second Edition. 156 pp. (1857). Crown 8vo. 5s. 6d.

Dynamics of a Particle. With numerous Examples. By P. G. TAIT, M.A. and W. J. STEELE, M.A. 304 pp. (1856). Crown 8vo. 10s. 6d.

A Treatise on Dynamics. By W. P. WILSON, M.A. 176 pp. (1850). 8vo. 9s. 6d.

Dynamics of a System of Rigid Bodies. With numerous Examples. By E. J. ROUTH, M.A. 336 pp. (1860). Crown 8vo. 10s. 6d.

Geometrical Treatise on Conic Sections. With a Collection of Examples. By W. H. DREW, M.A. 121 pp. (Second Edition, 1862). 4s. 6d.

Solutions to Problems contained in a Geometrical Treatise on Conic Sections. By W. H. DREW, M.A. (1862). 4s. 6d.

Elementary Treatise on Conic Sections and Algebraic Geometry. By G. H. PUCKLE, M.A. Second Edition. 264 pp. (1856). Crown 8vo. 7s. 6d.

Elementary Treatise on Trilinear Co-ordinates. By N. M. FERRERS, M.A. 154 pp. (1861). Crown 8vo. 6s. 6d.

A Treatise on Solid Geometry. By P. FROST, M.A. and J. WOLSTENHOLME, M.A. 8vo. [In the Press.

A Treatise on the Calculus of Finite Differences. By GEORGE BOOLE, D.C.L. 248 pp. (1840). Crown 8vo. 10s. 6d.

The Algebraical and Numerical Theory of Errors of Observations and the Combination of Observations. By the Astronomer Royal, G. B. AIRY, M.A. Pp. 103 (1861). 6s. 6d.

The Construction of Wrought Iron Bridges, embracing the Practical Application of the Principles of Mechanics to Wrought Iron Girder Work. By J. H. LATHAM, M.A. C.E. With numerous plates. Pp. 282 (1858). 15s.

Mathematical Tracts. On the Lunar and Planetary Theories, the Figure of the Earth, Precession and Nutation, the Calculus of Variations, and the Undulatory Theory of Optics. By the Astronomer-Royal, G. B. AIRY, M.A. Fourth Edition (1858), pp. 400. 15s.

An Elementary Treatise on the Planetary Theory. By C. H. H CHEYNE, B.A. Scholar of St. John's College, Cambridge. [Preparing.

A Treatise on Attractions, Laplace's Functions, and the Figure of the Earth. By J. H. PRATT, M.A. Second Edition. Crown 8vo. 126 pp. (1861). 6s. 6d.

An Elementary Treatise on Quaternions. By P. G. TAIT, M.A., Professor of Natural Philosophy in the University of Edinburgh. [Preparing.

Singular Properties of the Ellipsoid, and Associated Surfaces of the Ninth Degree. By the Rev. G. F. CHILDE, M.A. Mathematical Professor in the South African College. 8vo. 10s. 6d.

Collection of Mathematical Problems and Examples. With Answers. By H. A. MORGAN, M.A. Pp. 190 (1858). Crown 8vo. 6s. 6d.

Senate-House Mathematical Problems. With Solutions

1848-51. By FERRERS and JACKSON. 8vo. 15s. 6d.
1848-51. (Riders.) By JAMESON. 8vo. 7s. 6d.
1854. By WALTON and MACKENZIE. 8vo. 10s. 6d.
1857. By CAMPION and WALTON. 8vo. 8s. 6d.
1860. By ROUTH and WATSON. Crown 8vo. 7s. 6d.

Hellenica: a First Greek Reading-Book. Being a History of Greece, taken from Diodorus and Thucydides. By JOSIAH WRIGHT, M.A. Second Edition. Pp. 150 (1857). Fcp. 8vo. 3s. 6d.

Demosthenes on the Crown. With English Notes. By B. DRAKE, M.A. Second Edition, to which is prefixed Æschines against Ctesiphon. With English Notes. (1860.) Fcp. 8vo. 5s.

Juvenal. For Schools. With English Notes and an Index. By JOHN E. MAYOR, M.A. Pp. 464 (1853). Crown 8vo. 10s. 6d.

Cicero's Second Philippic. With English Notes. By JOHN E. B. MAYOR. Pp. 168 (1861). 5s.

Help to Latin Grammar; or, the Form and Use of Words in Latin. With Progressive Exercises. By JOSIAH WRIGHT, M.A. Pp. 175 (1855). Crown 8vo. 4s. 6d.

The Seven Kings of Rome. A First Latin Reading-Book. By JOSIAH WRIGHT, M.A. Second Edit. Pp. 138 (1857). Fcp. 8vo. 3s.

Vocabulary and Exercises on 'The Seven Kings.' By JOSIAH WRIGHT, M.A. Pp. 94 (1857). Fcp. 8vo. 2s. 6d.

First Latin Construing Book. By E. THRING, M.A. Pp. 104 (1855). Fcp. 8vo. 2s. 6d.

Sallust.—Catilina et Jugurtha. With English Notes. For Schools. By CHARLES MERIVALE, B.D. Second Edition. Pp. 172 (1858). Fcp. 8vo. 4s. 6d. Catilina and Jugurtha may be had separately, price 2s. 6d. each.

Æschylus.—The Eumenides. With English Notes and Translation. By B. DRAKE, M.A. Pp. 144 (1853). 8vo. 7s. 6d.

St. Paul's Epistle to the Romans. With Notes. By CHARLES JOHN VAUGHAN, D.D. (1861). Crown 8vo. 5s.

The Child's English Grammar. By E. THRING, M.A. Demy 18mo. New Edition (1857). 1s.

Elements of Grammar taught in English. By E. THRING, M.A. Third Edition. Pp. 136 (1860). Demy 18mo. 2s.

SPOTTISWOODE AND CO., PRINTERS, NEW-STREET SQUARE, LONDON

www.ingramcontent.com/pod-product-compliance
Lightning Source LLC
LaVergne TN
LVHW050519100826
845148LV00002B/382

* 9 7 8 1 4 2 5 5 2 0 4 2 7 *